Developments in the Analysis of Lipids

Developments in the Analysis of Lipids

Edited by

J. H. P. Tyman
Brunel University

M. H. Gordon
University of Reading

THE ROYAL SOCIETY OF CHEMISTRY

The Proceedings of a conference organised jointly by the SCI Oils and Fats
Group and the RSC Lipid Chemistry Group on Developments in the Analysis of
Fats and Other Lipids, held at The University of Newcastle upon Tyne, on
20–21 April 1993.

Special Publication No. 160

ISBN 0-85186-971-8

A catalogue record of this book is available from the British Library.

Published by The Royal Society of Chemistry,
Thomas Graham House, The Science Park, Milton Road,
Cambridge CB4 4WF, UK

Printed by Bookcraft (Bath) Ltd.

Preface

This book has its origins in a one-day meeting on lipid analysis organised jointly by the Lipid Group of the Perkin Division of the Royal Society of Chemistry and the Oils and Fats Group of the Society of Chemical Industry at the University of Newcastle upon Tyne in 1993. It is some years since a previous meeting on this topic was held at the University of Reading in 1988 arranged by the Lipid Group and the Biochemical Society.

Analysis has been said to be the basis of chemistry. This is particularly true of lipid chemistry which has led the way in many aspects-one has only to remember the classic origin of gas chromatography by Martin and colleagues in 1952 on the separation of fatty acids following a prediction in 1946 of its feasibility by the same author. Lipids represent the largest replenishable natural resource for direct and semi-synthetic use by mankind, and compositional analysis is of great interest and importance both to identify trace materials and to locate new and abundant sources of major components.

Nowadays there is a plethora of chromatographic and spectroscopic methods together with combined procedures and a wealth of excellent and invaluable technical literature emanating from commercial suppliers purveying first class equipment. However, both for the novice and the experienced analyst it is necessary to examine the origin and background of chromatographic and spectroscopic analytical techniques. In the last few years there have been very few meetings and only just recently a number of texts on lipid analysis, notably, 'Lipid Analysis', (ed. by R J Hamilton and S Hamilton), 'Advances in Lipid Methodology-one', 'Advances in Lipid Methodology- two', (ed. by W W Christie), 'Chromatography for the Analysis of Lipids', by E Hammond and 'Advances in Lipid Research' (ed. F Padley). This book gives material in some other aspects of the technology.

An analytical book should, in the present technological era, hope to have both an educative and an industrial value. This slender volume aims to present such a picture and to appeal to a wide audience. The authors are almost equally divided into those with industrial and academic backgrounds. Each is a well known practitioner in his field and often beyond. Almost inevitably some important topics could not be included such as an account of developments in the FTIR method and in the specialised subject of radiotracers.

Unfortunately it did not prove possible to obtain and include all the papers presented, at the original symposium, in particular a contribution on the analysis of 'other lipids', and to redress the balance of the title,

'Developments in the Analysis of Fats and Other Lipids', one of the editors has contributed a chapter on the analysis of phenolic lipids.

A chapter on supercritical fluid extraction and chromatography is presented by K D Bartle and A A Clifford. The emphasis is on the extractive aspect and the factors which govern selectivity and quantitative recovery in this technique which has a number of important industrial applications.

L G Blomberg and M Demibuker in their contribution discuss four different approaches for the quantitative analysis of triglycerides, including supercritical fluid chromatography, silver ion chromatography, the use of packed micro columns and miniaturised evaporative light scattering detection.

The requirement to have pure enantiomeric compounds for evaluation is well recognised and established in the pharmaceutical industry and this aspect is also important in lipid chemistry. W W Christie discusses the chromatographic analysis of chiral lipids with particular reference to the question of suitable derivatives for the resolution of diastereomers and to the development of novel stationary phases for HPLC with chiral moieties bonded chemically to inert supports.

A chapter concerning on-line LC-GC for the analysis of edible oils and fats is contributed by K Grob. The conventional method for the analysis of the characteristic minor components of oils and fats, of significance in their own right and in adulteration studies, is somewhat laborious and the on-line approach described aims to improve this by the use of transesterification and direct analysis by combined LC-GC.

HPLC and GC methods for the analysis and identification of 'symmetrical' triglycerides, particularly of the major natural material, cocoa butter, are described by E Hammond in his account which details an important subject in the confectionery industry.

Developments in the analysis of the industrial non-edible phenolic lipids such as technical cashew nut-shell liquid and oriental 'lac' are outlined by J H P Tyman.

Spectroscopic methods are of course widely and routinely used in lipid analysis; emphasis was directed in the initial part of this section of the symposium to less familiar techniques. F D Gunstone, of long standing in all aspects of lipid research, has contributed a chapter on the use of ^{13}C NMR for providing both qualitative and quantitative information on the structural analysis of pure lipidic compounds and its application to a wide variety of natural mixtures.

An important and perhaps neglected topic in lipid chemistry namely that of crystallisation of lipids and the colloidal structure of these complex systems has been studied by neutron scattering and interesting and new conclusions have been put forward in a chapter contributed by M J W Povey.

Mass spectrometry is invaluable in the analysis of lipids of all types and has revolutionised the whole field of study. In a comprehensive account R P Evershed has admirably summarised all recent techniques of importance to

the lipid chemist and clearly indicated the comprehensive
information which can be derived by a variety of different
analytical procedures.
J B Rossell gives an account of a vast amount of work
extending over many years on purity criteria in edible oils
and fats which involves the use of both GC and spectrometry
in the form of stable carbon isotope ratio measurement;
indeed his contribution extends into both halves of the
original symposium on chromatographic and spectroscopic
techniques and perhaps indicates that analysis is one very
large indivisible subject.
The editors wish to thank all the contributors for their
efforts in recording their expertise and giving a wide
choice of reading and bibliographic material of value in
lipid chemistry and probably beyond. The RSC are thanked
for their help in the publication of this work.

LONDON, SW14   JOHN TYMAN

                                            MICHAEL GORDON

Contents

Supercritical Fluid Extraction and Chromatography of Lipid Materials

K. D. Bartle and A. A. Clifford

SUPERCRITICAL FLUID GROUP, SCHOOL OF CHEMISTRY,
UNIVERSITY OF LEEDS, LEEDS LS2 9JT, UK

1. ADVANTAGES OF USING A SUPERCRITICAL FLUID FOR EXTRACTION AND CHROMATOGRAPHY

A supercritical fluid is a substance above its critical temperature and pressure. Above its critical temperature, it does not condense or evaporate to form a liquid or a gas, but is a fluid, with properties changing continuously from gas-like to liquid-like as the pressure increases. Table 1 shows the critical parameters of some compounds useful as supercritical fluids.

Table 1. Critical parameters of selected substances useful as supercritical fluids

	T_c (°C)	p_c (atm)	ρ_c (10^3kg m^{-3})
CO_2	31.3	72.9	0.47
N_2O	36.5	72.5	0.45
SF_6	45.5	37.1	0.74
NH_3	132.5	112.5	0.24
H_2O	374	227	0.34
$n\text{-}C_2H_{10}$	152	37.5	0.23
$n\text{-}C_5H_{12}$	197	53.3	0.23
Xe	16.6	53.4	1.10
CCl_2F_2	112	40.7	0.56
CHF_3	25.9	46.9	0.52

One compound, CO_2, has so far been the most widely used, because of its convenient critical temperature, cheapness and non-toxicity (of great importance in the food industry). Because the molecule is non-polar it is classified as a non-polar solvent, although it has some limited affinity with polar solutes because of its large molecular quadrupole. For the extraction and chromatography of polar molecules, therefore, it is common to add modifiers or entrainers, such as the lower alcohols, to CO_2, usually in small quantities. In such cases, it is important to be aware of the modifier-CO_2 phase diagram to ensure that the solvent is in one phase. For example for methanol-CO_2 at 50°C there is only one phase above 95 bar whatever the composition, but below this pressure, certain compositions will be liquid, some gaseous (supercritical) and some separate into both of these phases.[1] It should be mentioned that, for both pure fluids and mixtures, many of the advantages of a supercritical fluid are possessed by liquids which are just sub-critical, and these are used in industrial processes, for example in the extraction of hops. The term near-critical is used to describe both situations and is preferred by some people.

Supercritical fluid extraction (SFE) and supercritical fluid chromatography (SFC) take advantage of the fact that a supercritical fluid can have properties controllable by pressure. Table 2 shows some rather approximate typical values of important properties: density (this is related to solvating power), viscosity (related to flow rates) and diffusion coefficients (related to mass transfer within the fluid).

Table 2. Typical physical property values for gases, supercritical fluids and liquids

	Density (10^3kg m^{-3})	Viscosity (mPa s)	Self-diffusion coefficient $(10^4 \text{m}^2\text{s}^{-1})$
Gas 30°C, 1 atm	$0.6\text{-}2 \times 10^{-3}$	$1\text{-}3 \times 10^{-2}$	0.1-0.4
Supercritical fluid			
near $T_{c1}p_c$	0.2-0.6	$1\text{-}3 \times 10^{-2}$	0.7×10^{-3}
near $T_{c1}4p_c$	0.4-0.9	$3\text{-}9 \times 10^{-2}$	0.2×10^{-3}
Liquid 30°C, 1 atm	0.6-1.6	0.2-3	$0.2\text{-}2 \times 10^{-5}$

The principal advantage for SFE is that solubilities, and particularly the relative solubilities of two compounds can be controlled via both pressure and temperature, making extraction selective to some extent. Other advantages are the relatively easy removal of the solvent and the non-toxicity and cheapness of CO_2. The advantage for SFC is that the higher diffusion coefficient gives narrower peaks and better separation for reasonable analysis times, while at the same time the solvating effect of the fluid permits chromatography of thermo-labile and higher molar-mass compounds impossible to analyse by gas chromatography. The disadvantage of using a supercritical fluid is that high-pressure technology is involved. Although SFE and SFC are the two areas where supercritical fluids have been widely exploited, research into the use of these fluids in other areas, such as preparative SFC, chemical reactions, microcrystallisation, dyeing, painting and electrochemistry is proceeding. It must be finally said that SFE and SFC are applicable to particular problems, where existing techniques are experiencing problems and excessive euphoria as to their usefulness must be avoided.

2.   BASIC PRINCIPLES OF SFE

Extraction by a supercritical (or any) fluid is never complete in finite time, but may be considered to be successful under some conditions on the basis of either accuracy, in the case of analytical extraction, or economics, in the case of process extractions. For extraction to be successful three main criteria must be satisfied. The solute must, firstly, be sufficiently soluble in the supercritical fluid; secondly, the solute must be transported sufficiently rapidly, by diffusion or otherwise, from the interior of the matrix in which it is contained; and thirdly, interactions between the solute and the matrix, such as adsorption, must be overcome.

Solubility is the subject of sections 4,5, and 6, and only a few major points will be made here. Firstly, the solubility can be described either as the mole fraction of solute in the solution, x, or by the amount (as moles or mass) of solute per unit volume, S, at saturation. The relationships between these quantities are trivial and involve molar masses and the density of the solution, which, as the solutions are dilute, is often taken to be the density of the pure fluid.

Solubility of a substance in a supercritical fluid is contributed to by two factors: the volatility of the substance and the solvating effect of the supercritical fluid. The latter factor, as we shall see more quantitatively in the next section, is primarily a function of the density of the fluid. The solubility of a substance at constant temperature as a function of pressure has the schematic form of Figure 1. In section A-B, which is at very low pressures, which are not of concern in most extraction processes, x falls as the solute is diluted by the fluid. In section B-C there is a rapid rise in x at a so-called threshold pressure characteristic of the solute-fluid system, which is a pressure somewhat above the critical pressure of the fluid. This occurs because of the rapid rise in the density, and therefore solvating effect, of the fluid at around this pressure. The next two features may or may not occur for any particular system, especially as the pressure range of interest may be limited. A fall, shown as C-D may occur because, at higher pressure, repulsive forces may 'squeeze' the solute out of solution. And the rise shown as D-E only occurs for rather volatile solutes, if there is a critical line in the mixture phase at higher pressures. For many extraction processes, it is only the section of the curve B-C that is of interest.

At constant pressure and as a function of temperature, solubility has a minimum at a particular temperature. The initial fall occurs because, as the temperature rises, the density and solvating effect falls. However, as the temperature rises, the volatility of the solute also rises, and eventually this effect exceeds the effect of the falling solvation and the solubility rises.

The solubility of a substance is only a guide to its extractability. The solubility gives the concentration of solute in equilibrium with the pure solute. The presence of other materials in the matrix and in the fluid will effect the equilibrium concentration. The effect of the matrix will be, in general to reduce this concentration below the solubility. If the matrix can be considered to be an ideal liquid mixture, the equilibrium concentration of each component will be equal to its solubility multiplied by its mole fraction in the matrix. Most systems will be non-ideal to an unknown extent, and the only conclusion that can be drawn is that the equilibrium concentration will fall as the concentration in the matrix falls. The presence of other solutes in the fluid, in general, enhances the equilibrium concentration, [2] as the solutes usually have a stronger affinity to each other than to the solvent fluid. In addition to these effects, in any real extraction, because of the kinetic effect of diffusion out of the matrix, any solute will always be below its equilibrium concentration with the matrix, for extraction to be occurring. Nevertheless, in spite of the caveats, the extent of extraction of a solute after a given time, plotted as a function of

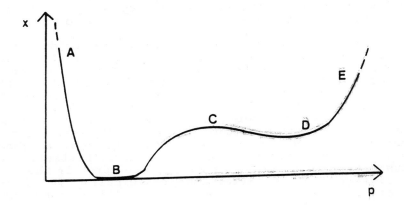

Figure 1. Features which can occur in solubility isotherms.

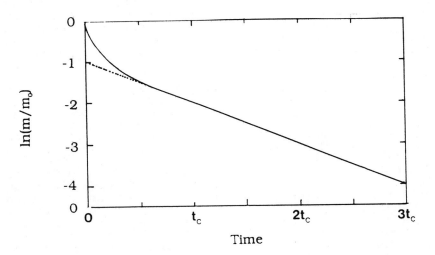

Figure 2. Plot of $\ln(m/m_0)$ versus time for SFE with little solubility limitation.

pressure at constant temperature (see Figure 4b), often follows the same shape as the corresponding solubility curve and has approximately the same threshold pressure.

We now consider the second effect, that of diffusion out of the matrix, and to illustrate this, in isolation from the solubility effect, we consider the extraction of a matrix in a flow of fluid, which is fast enough for the concentration of a particular solute to be well below its solubility limit. If the mass of solute in the matrix is m_0 initially and m after a given time, a plot of $\ln(m/m_0)$ versus time has the form given by Figure 2. It is characterised by a relatively rapid fall on to a linear portion, corresponding to an extraction 'tail'.[3] The physical explanation of the form of the curve is that the initial portion is extraction, principally out of the outer parts of the matrix particles, which establishes a smooth concentration profile across each particle, peaking at the centre and falling to zero at the surface. When this has happened, the extraction becomes an exponential decay.

The curve is characterised by two parameters: a characteristic time, t_c, and the intercept(I)of the linear portion, -1. The slope of the linear portion is $-1/t_c$ and the linear portion begins at approximately $0.5t_c$; t_c is related to the effective diffusion coefficient out of the matrix, D, and a mean dimension, L, (say the longest distance across)of a matrix particle by the equation

$$t_c = \frac{AL^2}{D} \tag{1}$$

A is a constant, dependent on particle shape (which will be varied) and the definition of L, in general will not be known, but it is of the order of 0.1. For some systems, like liquid droplets, D will be a true diffusion coefficient, but often it will be an effective diffusion coefficient representing such processes as migration between adsorption sites, diffusion out of pores or passage of the fluid into and out of a porous particle. Most measurements published for D are for small molecules in relatively mobile solvents[4] and D is of the order of 10^{-5} cm^2 s^{-1}. For systems of interest to the food industry D will be of between 1 (for oils) and 4 (for solids) orders of magnitude below this value. In most cases a value will not be available, although some prediction methods are possible.[5]

Equation (1) shows an direct square dependence on L and rationalises the commonsense rule that for rapid extraction matrix particles must be small. This may be achieved for solids by crushing or grinding and for liquids by coating on a finely divided substrate or spraying or mechanical agitation. For solid matrix particles with L of the order of 1 mm, typical values of t_c are between 10 and 100 minutes.

The value of I depends on the particle shape and size distribution (for the former in particular the surface to volume ratio) and also the distribution of solute within the matrix particles (i.e. whether the solute is primarily located near the surface or in the interior of the particle). For a model system of spheres of the same size, with uniform solute concentration, it is 0.5. For real systems values of *ca*. 2 are common and prediction of the values is not really possible. Thus usually values of t_c and I can only be obtained by experiment. A small-scale dynamic extraction followed by the application of an appropriate analytical technique, as described in a later section, is therefore an important preliminary study in designing a process or a routine quantitative analytical procedure. For the former, such a study will provide information on the time scale for extraction and on the expected efficieny of extraction after a particular time, information which can be then considered in conjunction with solubility considerations. An experimental method for such a study is given in section 3.

The information of Figure 2 is reproduced in Figure 3 in terms of percentage extraction versus time. As can be seen the majority is extracted in a time of $0.5t_c$; 92% for the example given, where I has been taken as equal to 2. Thereafter there is a long 'tail', corresponding to slow extraction: only a further 3% is extracted in the next period of $0.5t_c$. These data represent a situation where there is no limitation of the extraction rate by solubility considerations. In a real situation, therefore, the extraction rates will be less, depending on the nature of the extraction process used.

To illustrate the interaction between solubility and matrix-diffusion considerations, extraction of a solute from a stationary matrix in a cell by steady flow of the fluid is now discussed. Flow rates are now slow enough so that, at the beginning of the extraction, the concentration of the fluid in the solute is an appreciable fraction of its equilibrium concentration, which will vary with pressure. Figure 4a shows plots of the type of Figure 2, but now with solubility limitation, at a number of pressures, but constant flow rate, curve 1 being at the lowest pressure and curve 4 at the highest.

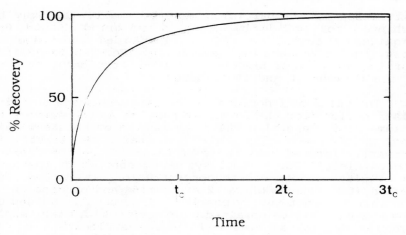

Figure 3. Plot of percentage extracted versus time for
SFE with little solubility limitation.

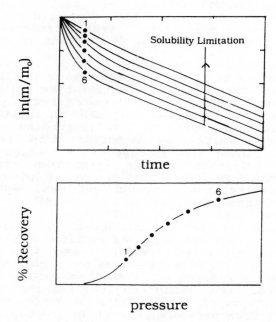

Figure 4.(a) Plots of $\ln(m/m_0)$ versus time at various
pressures and constant temperatures, with the highest
curve (1) at the lowest pressure and hence solubility.
(b) Percentage extracted after a given time, plotted
against pressure, with the points 1 - 6 corresponding to
the points on (a).

The effect of solubility limitation is to reduce the dramatic high rate at the beginning of the extraction and delay the onset of the linear portion, so that it moves upwards and to the right. However eventually solubility limitation will disappear and the slope will be that of the original curve, as in Figure 2. The extent of this effect will increase as the pressure falls and the solubility decreases. If data are taken at a particular time, (this effect is most obvious at the beginning of the extraction) and plotted as a function of pressure, then Figure 4b is obtained, which is a typical pressure-threshold curve and similar in shape to the solubility curve for the solute. A knowledge of solubility from published results, estimation by prediction, or measurement is therefore important.

The third problem, that of interaction between the solute and the matrix, is the least well understood. It can result in only a fraction of the solute being extracted, even after a long time. Methods of dealing with it include the use of higher temperatures and of modifiers. Modifiers are believed to be as important in releasing solutes from matrix interactions as they are in increasing solubility in the fluid.

3. EXPERIMENTAL DETERMINATION OF KINETIC EXTRACTION DATA

To obtain kinetic extraction data, suitable for plotting as in Figures 3 and 5, continuous extraction at constant flow rate, temperature and pressure is carried out. The extract is collected for a series of time intervals, e.g. by passing it through a suitable solvent, and then analysed by any appropriate method. Extraction is continued until it is complete to within 99% or better. This is done to obtain the value of m_0 to within 1% and thus values of $\ln(m/m_0)$ plotted below are fairly reliable down to a value of *ca.* -3, corresponding to 95% extraction. The method used by Hawthorne and his coworkers, [6,7] for extractions with pure or modified CO_2, is briefly described below, as an example.

The CO_2, is supplied at constant pressure using a suitably programmed syringe pump, which has a cooled pump head. Modifier, if used, is supplied at constant flow rate by a second pump. The supercritical fluid then passes to extraction cells with internal volumes of *ca.* 1 cm³ which have been previously filled with weighed test samples and placed in a thermostatted tube heater to maintain constant temperature. The flow rate is controlled and kept constant by a capillary restrictor, and this is monitored, in terms of liquid CO_2 volume flow, at the pump. For example, using pure CO_2 at 50°C and 400 atmospheres, a flow rate of *ca.* 0.45 cm^3 min^{-1} (measured as liquid CO_2 at the pump) is obtained with a 10 cm length of fused silica tubing of 25 μm internal diameter at the extraction cell outlet. The compounds extracted from each sample are then collected for several time intervals by purging the extraction cell effluent into *ca.* 3 cm^3 of an appropriate solvent.[7] An internal standard is then added to each timed extract, and the concentrations of each target species determined using, for example, capillary gas chromatography. Extraction was continued until there was a very small or no further detectable extract and it was estimated that in excess of 99% had been extracted.

4.   METHODS FOR OBTAINING SOLUBILITY

The various experimental methods used to obtain supercritical fluid solubilities can be classified in two different ways. The first relates to the way in which that saturated solution is obtained, which can be static or dynamic, i.e. in a closed cell or in a flow system. The second classification describes how this solution is analysed and the methods used can be grouped into four categories, gravimetric, spectrometric, chromatographic and miscellaneous. There is some correspondence between these two classifications. To produce enough material for accurate gravimetric analysis, a flow system is normally used. Reviews of experimental methods for supercritical solubilities have appeared recently[8,9].

Of the techniques described, the gravimetric method is most widely used, with a few research groups, who having established a particular procedure and given data in one publication, following up with subsequent publications using essentially the same apparatus. Chromatographic analysis is the second most popular technique. Other methods tend to be limited to a single paper, describing the development of the experimental procedure and giving a body of data so obtained. A fair assessment of the present situation is therefore that the gravimetric method with its variants and to a lesser extent chromatographic analysis are the present established and reliable procedures and these two methods are described below. Also described is the simpler and faster, but less direct and sometimes qualitative, method of using chromatographic retention for obtaining solubilities.

Most of the gravimetric methods used have the same basic structure. The main motivation has been the need for solubility data for extraction processes and the philosophy of the methods reflects this interest. Briefly, these methods involve the production of a saturated solution by passing the supercritical fluid over the solute in an extraction cell, dropping the pressure to precipitate the solute and weighing it. The experimental procedures were developed by Eckert, Paulaitis and Reid and their coworkers in the late 1970s. [10-12]

A schematic diagram of the basic system is shown in Figure 5. The CO_2 is pumped, as a gas by a compressor or as a liquid by a pump with a cooled head into a thermostat, where it first passes through a preheating coil. It then passes into an extraction or equilibrium cell where a saturated solution is formed. The solute is usually dispersed in the cells, e.g. by coating it on to sand particles, and filters are positioned at the ends of the cells to prevent the entrainment of undissolved solute. The solution is then dropped to atmospheric pressure via a restrictor or valve, such as a back-regulator, which is heated to prevent the solute being lost in the valve and clogging it. The gas containing, finely divided solute precipitate, passes to a trapping system, which can vary in type (e.g. U-tubes are often used), or complexity (e.g. switching circuits between traps are sometimes described). More than one trap in series is often used and methods for ensuring complete trapping, such as cooling and/or packing with absorbent, are typically employed. Finally the CO_2 passes through some kind of flow meter. The pressure in the system, i.e. that of the experiment and the flow-rate are controlled by the pump and the valve. In the most straighforward system, a back-pressure regulator would control the pressure and the pump the flow-rate.

A typical experiment is to set the flow and allow the system to reach a steady state and then switch in a weighed trap or traps for a measured period, during which the rate of flow of CO_2 is monitored. After reweighing the trap(s) and calculating the total mass of CO_2 passing in the period, the solubility is obtained directly in terms of mole fraction. Usually the errors in the experiment are quoted as 3-5% for solubilities obtained by this method. An important consideration is that equilibrium has been reached in the extraction cell and different ways are used to ensure and check this including varying the flow rate. [13] The other important test is that all the precipitated solute has been collected after depressurising. This can be done by using two or more successive traps and weighing both to show that the great majority of solute is collected in the first trap. [14] Problems can arise with precipitation inside the

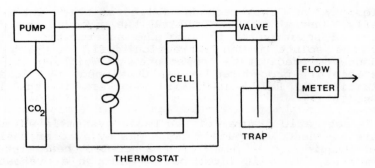

Figure 5. Schematic diagram of an apparatus for the gravimetric determination of solubility.

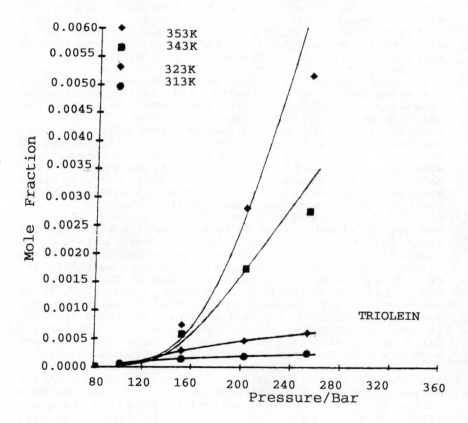

Figure 6. Published experimental values [24] for the solubility of triolein in CO_2.

pressure-reducing valve, which can be overcome by washing out the valve after an experiment. [15]

Most chromatographic methods used are modifications of the gravimetric method shown in Figure 5. In one type of modification, the solute is precipitated as before in a trap, perhaps containing a solvent. The solute is then washed out of the trap, made up to volume and analysed by any suitable chromatographic method. In some cases the solute is precipitated directly on to a chromatographic plate. Gas chromatography (GC), [16] thin layer chromatography (TLC), [17] high performance liquid chromatography (HPLC), [18] including size exclusion chromatography (SEC sometimes called GPC) [19] are used. After calibration and analysis, followed by calculation of the mass of CO_2 passing during the sampling period as before, the method gives solubilities in terms of mole fractions directly. Accuracy comparable with the gravimetric method has been obtained, but some variants of the method, e.g. using direct precipitation on to a TLC plate, are intended only to give a rough indication of solubilities.

A second type of modification of the apparatus of Figure 5 for chromatographic analysis, consists of removing the trap and flow-meter and inserting a sampling valve between the extraction cell and the valve or back-pressure regulator. This is commonly a multiport chromatographic sampling valve, with a sample loop. Using this device a fixed volume of the saturated solution is sampled and injected directly into a chromatograph (or alternatively let down to atmospheric pressure, the solute dissolved in a solvent and then injected into a chromatograph) and analysis carried out by GC, [20] HPLC [21] or supercritical fluid chromatography (SFC). [22] Because only small samples are needed, the extraction section of the apparatus can be small-scale and the commercially available supercritical fluid extraction (SFE) systems designed for analytical applications, [23] or even a small static system, [24] can be used. The apparatus is calibrated by loading a liquid solution of the same solute of known concentration into the same sample loop. The solubility results are obtained directly in terms of concentration rather than mole fraction. Careful use of this method can also give results of accuracy comparable with the gravimetric method.

A completely different way of obtaining information on solubilities, which is less direct and less accurate than the best direct experiments, is to use chromatographic retention. The degree of retention of a solute in chromatography, as measured by the capacity

factor, k', is at least qualitatively inversely related
to the solvating power of the mobile phase for that
solute: the more soluble it is in the mobile phase, the
less it will be retained. This method has been used to
obtain qualitatively the dependence of solubility on
pressure and to obtain 'threshold pressures' for
extraction. In some cases solubility data has been
given[25,26].

Furthermore, it can be shown experimentally, that in
some situations at constant temperature and for a
particular column, the capacity factor for a particular
solute is fairly accurately inversely proportional to its
solubility in the mobile phase.

Such plots are not obtained for all systems and it
is necessary to check representative systems. However,
the relationship can provide a relatively rapid way of
generating large amounts of solubility data, if the
proportionality constant is known. This constant may be
obtained by a number of methods, including liquid
solubility measurements and relatively few supercritical
fluid measurements made by 'conventional' methods, and
solubility data have been published obtained in this way
[28].

5.   SOLUBILITY MEASUREMENTS AND THEIR CORRELATION

In the context of lipids, the published solubility
measurements reviewed here are restricted to long-chain
fatty acids, glycerides, steroids and some pesticides.
Other substances, which may be of interest in this area,
may be found by reference to a recent comprehensive
review of solubilities [9]. Also this section is mainly
restricted to the solubilities of pure compounds, which
have been published in tabular form, rather than
graphically, implying lower absolute accuracy. Details of
some more qualitative solubility measurements are
reported in the last section on industrial SFE. The
bibliography of solubilities of interest here gives the
compounds, temperature and pressure ranges, the
experimental method used and a reference to the original
publication. All these solubilities are in CO_2 and most
of the discussion in this and the next section refers to
this fluid. Examples of these data, the solubilities of
triolein,[24] are shown in Figure 6. These illustrate a
point, important in the discussion of industrial
extraction below, that at higher temperatures, where the
solute is more volatile, the solubilities exhibit the
feature D-E of Figure 1, i.e. are rising steeply with
pressure. This feature is not present at low
temperatures.

Correlation of supercritical fluid solubility data is not straightforward. All the features shown in Figure 1 can be reproduced qualitatively by any equation of state. For quantitative fitting more refined equations of state are useful in certain regions, and of these the Peng-Robinson[29] has been the most widely used. However, even this equation is not successful in fitting all the data at all pressures and temperatures. A further problem is that the parameters necessary for using the equation of state, such as the critical temperature and pressure of the solute and its vapour pressure and acentric factor, are not always available. This problem has been discussed in a paper by Johnston *et al* [30] who come to the conclusion that a cruder empirical correlation with density is the best available route for most compounds, and one method of this type is outlined later.

However, if a reasonable amount of data is available for a compound of interest and the other parameters are available, then the Peng-Robinson equation of state may be used to correlate data over a limited range of pressure and temperature. This will allow interpolation and limited extrapolation of the experimental data for design purposes. The solute-fluid interaction parameter, d, a fine-tuning parameter which is not known in advance, and obtained by fitting the equations to the experimental data, a procedure which also makes up for the approximations in the equations. [31] The parameter, d, is in principle temperature independent and so a value can be obtained by fitting all the data; a procedure which also makes up for the approximations in the equations. However, if isotherms are fitted separately, d is sometimes found to vary with temperature. This is especially the case when the solute is a solid which forms a liquid phase under certain conditions of temperature and pressure.

A much simpler correlation is to fit the solubility in terms of mole fraction, x, at constant temperature as a function of pressure to the following equation;

$$\ln\left(\frac{xp}{p_{ref}}\right) = A' + B\ (\varrho - \varrho_{ref});\qquad(2)$$

where p is the pressure, ϱ the density of the solution (approximately the density of the pure supercritical fluid), A' and B are constants at constant temperature and p_{ref} is a reference pressure, conveniently 1 bar. ϱ_{ref} is a reference density, chosen to be central to the density range of all the solubility data. A good value for pure and modified CO_2 would be 700 kg m^{-3} and the

following discussion assumes this value. The reason for
the choice of ϱ_{ref} is to make the values of A' obtained
much less sensitive to experimental error in the data and
easier to correlate between different sets of data and
different temperatures. A full discussion of this
correlation method has been given.[9] It is found that
better correlations are obtained if data below 100 bar
are not included in the fits and this has been done in
the examples given below.

6.   SOLUBILITY PREDICTIONS

For a solute for which there is a large volume of
published solubility data and for which critical
parameters, vapour pressure data and hence an acentric
factor is available, the best method for prediction of
solubility at any pressure and temperature would be to
correlate the data using an equation of state with
adjustable parameter(s) in the temperature-pressure
region of interest, as described using the Peng-Robinson
equation in the last section.

   For a solute for which there are solubility data
available, but not enough information, to carry out an
equation-of-state correlation, less accurate estimates of
the solubility may be obtained using the correlation of
equation 2 and the A' and B values obtained from the
experimental data. The simplest method is to assume that
B is constant and to correct A' for temperature using H_v.
More sophisticated variants of this procedure have been
described.[9]

   For solutes for which no experimental data are
available, approximate predictions for solubilities in
CO_2, which can be used as first estimates for design
purposes, can be obtained using the Peng-Robinson
equation. Usually estimates of the necessary parameters
and vapour pressures will have to be made by standard
methods,[5] and the interaction parameter, d, can be
estimated using the following equation:[33]

$$d = 0.51\,C\,(\omega_2 - \omega_1)\left(\frac{V_{c_2}}{V_{c_1}}\right)\left(\frac{p_{c_2}}{p_{c_1}}\right)^2 \qquad (3)$$

   In this equation C is a constant which is 0.5 for
molecules containing groups which may cause them to
dimerise in the gas phase, e.g. -OH or -COOH groups, but
is unity for other compounds. ω_i, V_c and p_c are the
acentric factors; critical molar volumes and pressures

respectively of CO_2 (subscript 1) and the solute (subscript 2). Critical volumes are generally not tabulated and these have been obtained from ref.[5]

$$V_{c,i} = \frac{(0.2918 - 0.0928\omega_i) RT_c}{P_{c,i}}$$ (4)

The equation was obtained by fitting data for some 20 compounds for which good solubilities and other parameters were available to the Peng-Robinson equation to obtain values of d. These values of d are plotted in figure 10 against the group of parameters on the left-hand side of equation (4) and it can be seen that there is a reasonable, if not perfect, correlation, with a line passing through the origin with a slope of 0.51.

7.   INDUSTRIAL SFE

Industrial supercritical extraction processes are now well established and textbooks are available on their chemical engineering aspects.[8,34,35] The technique is used on a large scale, for example using CO_2 for coffee decaffeination and hop extraction, and on a smaller scale, for the extraction of high-value natural products, such as perfumes.[36] More than 200 U.S. patents for supercritical extraction processes have been registered,[8] and a large number outside the U.S. mainly in Germany. A process for the extraction and fractionation of edible oils using supercritical propane, known as the Solexol process was developed by the M.W. Kellogg Co. in 1947.[38,39] This process is no longer in use, and at present it is CO_2 that is under consideration. Because of company confidentiality it is difficult to know to what extent industrial supercritical fluid extraction processes for lipids are being developed to the commercial stage. It would be generally known if a process of this type was commercially sucessful and was being carried out on a very large scale, like the coffee decaffeination mentioned above, and this does not appear to be the case at the present time. However, a number of patents and scientific publications has appeared indicating considerable interest in possible processes, and these have been reviewed in 1986.[40] Work is described below in five categories: lower pressure bulk extraction; bulk extraction at higher temperatures and pressures; fractionation during bulk extraction; modification of existing products; and obtaining high value components from lipid products.

In the former category a large number of studies have

been made on the extraction of various lipid-bearing
materials on a laboratory scale.[40] Many of these
resulted in qualitative or semi-quantitative estimates of
the solubilities of total oil or oil components. Recent
studies, all involving pure CO_2, have been made on the
solubility of pure and mixed triglycerides in
supercritical CO_2,[41] on equilibria involved in the
extraction of rapeseed oil,[42] on the use of adsorbents
in the SFE of vegetable oils,[43] and on the extraction
of cholesterol-free lipids from phytoplankton [44] by the
same fluid. Prominent amongst the earlier studies were
those of Stahl and coworkers,[45-47] and much of the
research has been carried out in Germany and also in the
U.S.A. Most investigations envisage the use of lower
pressures and temperatures with the extract solutions
obtained containing only a few percent of lipid extract.
The investigators see the advantage of CO_2 over the
presently used organic solvents such as hexane in its
non-toxicity, safety, ease of separation and cheapness.
Another advantage of using SFE as opposed to hexane
extraction is that the extract contains a smaller
proportion of undesirable components such as
phospholipids and pigments and 'degumming' of the product
can be avoided.

Many of the lipids to be extracted are sufficiently
volatile at temperatures of 60-80°C for their solubility
curves versus pressure to exhibit the steep rises at
higher pressures illustrated schematically by the section
D-E in the curve on Figure 1. This has been shown
experimentally in measurements on individual components,
for example in the data on triolein in Figure 6. It has
also been shown to be an effect for the extraction of
total oil by Stahl[46,47] and Friedrich and Pryde.[48] In
1984 a U.S. patent (4466923) was granted to Friedrich
which describes conditions for the replacement of hexane
by supercritical CO_2 for the large-scale extraction of
oilseeds, such as soy, cottonseed, sunflower, safflower
peanut and linseed; cereal components, such as corn germ;
and animal by-products, such as suet and offal. Although
this patent was granted, the behaviour of solubility on
which it is based is not unexpected and the experimental
evidence was in the public domain[46] at the time of the
patent application. The conditions suggested in the
patent are to use pure CO_2, temperatures of 60-80°C and
pressures above 550 bar, perhaps as high as 1200 bar. By
using the conditions suggested the patent author claims
that the weight percent of oil in the extracted solution
could be as high as 40%. The studies described in this
paragraph can be summarized as advocating a high
temperature and pressure route, which would have
increased technological costs, but the advantage of using

less solvent and a smaller volume throughput for a given yield.

Another advantage of using a supercritical fluid is that separation of the product into a number of fractions can be incorporated into a bulk extraction of the type described above.[49] This can be done either by increasing the pressure in stages during a batch extraction or reducing the pressure in stages and separating the precipitated fraction following total extraction in a batch, continuous or semi-continuous process. Fatty oils, for example can be separated into fractions which are rich in flavour and free fatty acids; rich in mono- and diglycerides; mostly triglycerides; and rich in waxes and pigments, by increasing the pressure in stages of 50-100 bar at 60°C. Similar fractionation of vegetable oils from soybean, rapeseed and sunflower seeds can be achieved.[46]

These bulk processes are envisaged on a scale of millions of tons per annum, but there is also interest in smaller scale processes designed to modify the physical properties, flavour and health status of fatty materials. The former include hardening vegetable oil products to make them suitable for confectionary use, which at present is done by organic solvent extraction, and the refining of vegetable oils.[50] The latter include the removal of fatty acids from vegetable oils,[51] from olive oil[52] and cheese[53] to improve or modify their flavour and of steroids like cholesterol from fats, for health reasons. This leads to the final category of SFE in relation to lipids, where natural lipid products are the source of compounds for medicinal use. This was the motivation for the study of the CO_2 solubility of a number of steroids,[54]. SFE has been used in the preparation of eicosapentenoic acid derivatives[55,56] and other medically active acids[57] from fish oils and the separation of lecithin from soybean oil has been described.[58]

8.   ANALYTICAL SFE

The use of SFE as a first stage in an analysis is rapidly gaining importance and instrument manufacturers have responded with the production of commercial systems in the last few years.

It has advantages in speed and pressure-controlled selectivity over solvent extraction and it can be made 'on-line' to subsequent analytical steps. There is a very recent review (1990) of this developing field,[59] which gives the current situation and detailed practical advice. In an off-line system a typical procedure is that described in Section 3, with a commercial system making for easier operation. For an on-line system, for example with SFE-SFC, extraction may be carried out by flowing the fluid through a thermostatted extraction cell at a particular pressure for a given time, letting the pressure drop through a restrictor, when the extract is deposited on the beginning of the column. The extraction is then stopped and chromatographic analysis carried out. A further extraction and analysis may then be carried out on the same sample to investigate remaining analytes.

Because of the solubility of most fat and oil components in CO_2, SFE, using the pure gas is highly applicable to lipid systems. Off-line SFE has been used to extract fat from meats [60] and on-line SFE-SFC to analyse fatty acids and glycerides from butter, cheese and oils.[61] Although triglycerides are very soluble under some conditions, selective extractions of organochlorine pesticides from fats have been obtained sufficiently fat free to permit their GC analysis, by extraction at 40°C and 120 bar.[62,63] In many cases, however, lipids must be removed from extracts before analysis of minor components by solid phase extraction.

There are still problems in using these systems for quantitative analysis, which are the result of the long extraction tail, visible in Figure 3. Extraction in an initial period may extract 70% and in a subsequent extraction 7% only obtained. There may be a temptation for the analyst to conclude that a further extraction will drop in the same proportion and little extract remains. This is not the case, more than 20% remains and substantial yields in further extractions are obtained. When samples are almost constant in size, shape and composition it may be possible to extract under constant conditions and apply a factor in excess of unity (obtained from a lengthy extraction study) to the results. However, this has not been found easy or successful. No doubt these problems will find solutions soon. One possibility, where extraction is continuous, is to use the exponential tail to extrapolate the results to complete extraction and obtain quantitative analytical information in a shorter time than would be required for exhaustive extraction.[3] If extraction is carried out at least as long as the initial non-linear period to obtain an extracted mass m_1, followed by extraction over two subsequent equal time periods to obtain masses m_2 and m_3, then it can be readily shown the total mass in the sample, m_0, is given by

$$m_o = m_1 + \frac{m_2^2}{m_2 - m_3} \qquad (5)$$

Preliminary tests of the application of equation (5) to quantitation in analytical extractions is encouraging, and programmes of work to establish its usefulness are being carried out. If extraction is interrupted, however, as in the SFE-SFC on-line system described above, this equation does not apply. Diffusion within the sample continues to occur between extractions, and other mathematical solutions must be sought.

9.   SUPERCRITICAL FLUID CHROMATOGRAPHY

SFC separation of high molecular mass solutes occurs at temperatures well below those at which thermal decomposition occurs [64,65]. Because of the low viscosity of supercritical fluids, capillary columns may be employed, with consequent high resolution. The facility of operation of SFC with CO_2 as mobile phase means that the universal flame ionization detector (FID) as well as a range of GC and HPLC detectors may be used [66, 67]. Coupling of SFC to Fourier-transform IR (FTIR) [68] and mass spectrometric (MS) [69] detectors is also well advanced.

Figure 7 is a block diagram of the apparatus required for SFC. The mobile phase, most usually CO_2 either alone or with a small added concentration of a modifier, such as an alcohol or ether, is pressurized as liquid by a syringe pump. The liquid is delivered via an injection valve to the analytical column, which can be a conventional HPLC, microbore HPLC, packed capillary or fused-silica capillary column. The column is contained in an oven heated above the critical temperature of the mobile phase. A pressure restrictor at the end of the column ensures supercritical conditions in the column, and is installed either within the jet of an FID or in-line after a flow-cell detector such as a UV absorbance detector.

A major advantage of SFC is compatibility with both GC and HPLC detectors. Flame detectors (mainly FID, but also nitrogen and phosphorus thermionic, and flame-photometric sulphur selective detectors) have been interfaced [66] to the SFC column via a capillary or frit restrictor which allows decompression of the fluid as it enters the detector. HPLC detectors [67] (UV and

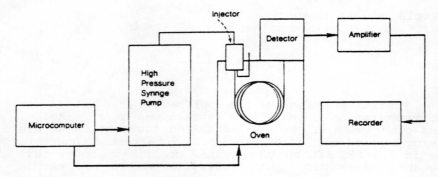

Figure 7. Schematic diagram of SFC instrumentation.

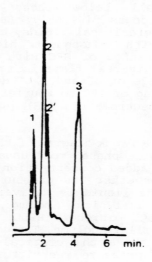

Figure 8. Separation of glycerides by SFC on a packed column.   1: triglycerides; 2: diglycerides; 3: monoglycerides. Column; 25 cm x 4 mm i.d. packed with 5-μm silica particles. Mobile phase: methanol (7.5%) modified CO_2 at 50°C. Detection: UV at 290 nm. (From Ref. [96], with permission.)

fluorescence) are necessary when the CO_2 contains an organic modifier; the restrictor is positioned after the flow cell. Alternative detectors for analytes without chromophores include light-scattering and ion-mobility devices.

FTIR detection after SFC may be carried out [68] by means of a flow-cell [70] in which the IR beam passes through the column effluent; mobile phase absorptions may interfere with detection of analyte, but a total absorption chromatogram may nevertheless be reconstructed in addition to the recording of spectra of individual peaks for identification. A solvent-elimination interface for FTIR detection [71] is also commonly applied; here the column effluent is sprayed onto an IR transparent plate from which the mobile phase is evaporated to leave a spot of analyte, later analysed by means of an FTIR microscope.

The interfacing of SFC to mass spectrometers has been energetically pursued [69]. The restrictor from a capillary column may be directly introduced into the ion source, although a further pumping stage before the ion source may be necessary for packed columns with higher gaseous flow rates [72].

10. APPLICATIONS OF SUPERCRITICAL FLUID CHROMATOGRAPHY TO LIPIDS AND RELATED COMPOUNDS

SFC is the most suitable technique for the separation of reactive, thermally labile and non-volatile compounds, among which groups are found many lipid-related compounds. Thus, free fatty acids, sterols etc. are most usually derivatised to improve volatility for GC [90, 91] with consequent increase in analysis time. Triglycerides are commonly analysed by HPLC [66], but with difficulties consequent on variable detector response; GC can be employed, but there is always a suspicion that decomposition occurs at the high column temperatures necessary.

Packed Column Supercritical Fluid Chromatography

Rawdon and Norris [75] used reverse-phase HPLC columns (ODS bonded silica) in the separation of oleic acid, mono-, di- and triglycerides, and soya and 'salad oil' triglycerides; the mobile phase was CO_2 modified with methanol with detection by UV absorption at 205 nm. While elution was very rapid (of the order of 2 min) resolution was limited. Perrin and Prevot [76] used adsorption mode

SFC with a silica column and a methanol-modified CO_2 mobile phase. Rapid separation (e.g. Figure 8) of mono-, di- and triglycerides was observed, with light-scattering, UV and FTIR detection; reduction of the methanol content of the mobile phase allowed a partial separation between 1,2- and 1,3-diglycerides. On an ODS column, further resolution of the individual bands (e.g. Figure 9) was possible, with separation of sunflower oil triglycerides on the basis of double bond number.

Underivatized fatty acids have been separated by SFC on a variety of packings (porous polymer [77], 78; ODS [79], etc.); modification of the CO_2 was necessary for satisfactory chromatography on the ODS (silica based) packings, but not for the porous polymer. Modification of CO_2 with ethanol and dichloromethane gave improved separation of tocopherols with good resolution of α and β isomers on a silica column [76].

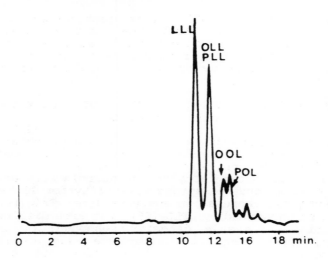

Figure 9. Separation by packed column SFC of triglycerides of sunflower oil. L: linoleic; O: oleic; P: palmitic. Column: 25 cm x 4 mm i.d. packed with 4-μm ODS modified silica particles. Mobile phase: methanol (0.7%) modified CO_2 at 40°C. Detection: UV at 290 nm. (From Ref. [76] with permission.)

A polar cyanopropyl silica column separated [80] saturated from unsaturated triglycerides with partial resolution of the latter, presumably by a dipole-induced dipole interaction. Similarly partial separations were obtained for SFC on columns designed for olefin separation in HPLC with Ag^+ ions loaded onto silica [80].

Capillary Column Supercritical Fluid Chromatography

Improved SFC resolution for mixtures of lipids and related compounds has been achieved on capillary columns. Early analyses by Chester [81] and White and Houck [82] demonstrated resolution similar to that available in capillary GC, but at much lower temperatures. The solubility of many lipids in CO_2 is sufficient to permit chromatography without modification of the mobile phase, although very highly polar compounds may require the addition of modifier; thus Raynor et al. were able to elute ecdysteroids with up to four hydroxyl substituents from a capillary column with CO_2 alone [83], but ecdysones with up to seven -OH groups were successfully chromatographed [83] (on a packed column) with methanol modified CO_2 as mobile phase (Figure 10). Kuei et al. preferred [84] to increase CO_2 solubility by analyte derivatisation, glycosphingolipids with molecular weights up to 2000 were analysed by capillary SFC with CO_2 mobile phase after permethylation. Steroids from physiological fluids were profiled by capillary SFC with phosphorus-thermionic detection of thiophosphinic ester derivatives [85].

Choice of the appropriate stationary phase is vital in lipid analysis by capillary SFC. Non-polar phases for capillary SFC, such as cross-linked methylpolysiloxanes [76, 81] allow separation of mono-, di- and triglycerides on the basis of carbon number (Figure 11). Excellent quantitation and retention time repeatability have been demonstrated [86] e.g. Table (3). More polar phases (phenylsiloxane, cyanopropylsiloxane and, particularly, Carbowax 20 M) yield further separation [87, 88] according to the degree of unsaturation. Thus Figure 12 illustrates the resolution of soya bean oil triglycerides. It is noteworthy that pressure or density programming of the mobile phase is necessary in all such analyses if results are to be compared with those from HPLC and GC. An SFC retention index system for non-polar phases has been advocated [89] with a linear pressure programme, related to the carbon numbers of the triglycerides. Triglyceride compositions for butter fat (Figure 13) and fish, rapeseed and soya bean oil were readily obtained by this procedure (e.g. Table 3). In

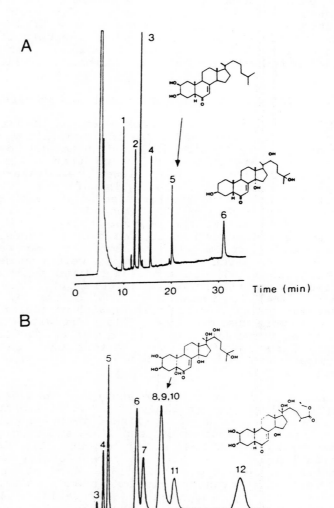

Figure 10.   Separation of ecdysteroids by SFC on (A)
capillary column (10 m x 50 μm i.d. coated with 0.25 μm
thick film of cross-linked cyanopropylmethylsiloxane).
Mobile phase: CO_2 density programme from 0.4 to 0.71 g
cm^{-3} at 0.015 g cm^{-3} min^{-1} after 5 min isoconfertic.
Temperature 120°C.   Detection: FID and (B) a packed
column (25 cm x 4.6 mm i.d. packed with 5 μm cyanopropyl
silica particles).   Mobile phase: methanol (10%)
modified CO_2. Detection UV at 215 nm.   (From Ref. [83],
with permission.)

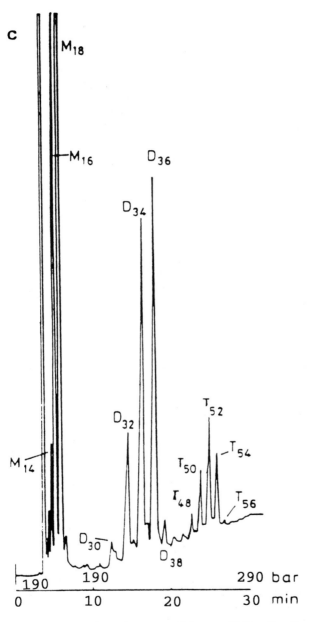

Figure 11. Separation by capillary SFC of glycerides. M: monoglycerides; D: diglycerides; T: triglycerides; subscript denotes carbon number. Column: 10 m x 100 μm i.d. coated with SE-54. Mobile phase: CO_2 pressure programme from 190 to 290 bar at 0.33 bar min^{-1}. Temperature: 170°C. Detection: FID. (From Ref. [87], with permission.)

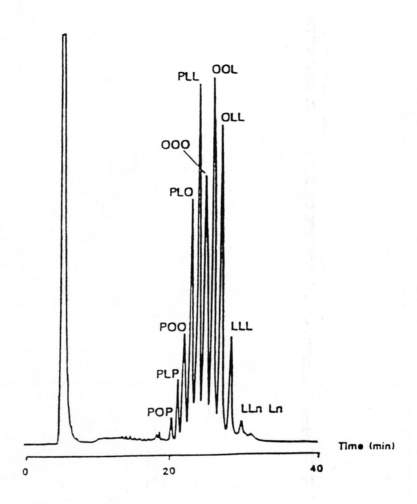

Figure 12. Separation by capilary SFC of triglycerides of soybean oil. P: palmitic; O: oleic; S: stearic; L: linoleic; Ln: linolenic. Column: 10 m x 50 μm coated with a 0.25 μm thick film of cross-linked Carbowax 20M. Mobile phase: CO_2 density programmed from 0.25 to 0.43 g cm^{-3} at 0.005 g cm^{-3} min^{-1}. Temperature: 200°C: Detection: FID. (From Ref.[88], with permission.)

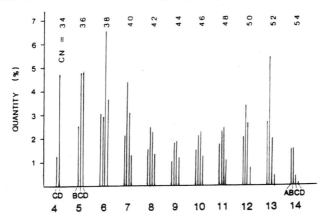

Figure 13.　Distribution of triglycerides in butter fat according to carbon number (CN) and degree of unsaturated (A: 3; B: 2; C: 1; D: 0)double bonds, determined by capillary SFC with flame ionization and secondary ion mass spectrometric detection.　Column:　5 m x 5 μm i.d. coated with a 0.2 μm film of DB-5 (95% dimethyl 5% diphenyl siloxane) Mobile phase: CO_2 pressure programmed from 13.8 to 27.6 MPa in 26 min and then to 34.5 MPa in 17 min after 2 min isobaric.　Temperature: 150°C.　(From Ref.[90], with permission.)

Table 3.　Repeatability of retention data and quantitation in the SFC analysis of triglycerides.

	Quantitation				Retention time (min)			
Run	$\%T_{36}$	$\%T_{42}$	$\%T_{48}$	$\%T_{54}$	$\%T_{R36}$	$\%T_{R42}$	$\%T_{R48}$	$\%T_{R54}$
1	20.76	24.89	26.03	28.32	16.41	20.84	24.44	27.55
2	20.71	24.55	26.41	28.32	16.39	20.83	24.44	27.54
3	20.90	24.84	26.36	27.89	16.41	20.86	24.44	27.54
4	20.63	24.73	26.30	28.34	16.45	20.86	24.42	27.53
X	20.75	24.75	26.27	28.22	16.41	20.85	24.43	27.54
σ	0.113	0.151	0.169	0.218	0.025	0.010	0.01	0.008
$\sigma_{rel}\%$	0.547	0.609	0.645	0.774	0.153	0 072	0.041	0.030

These data were measured on a 10 m x 100 μm SE-54 column Column temperature: 150°C isothermal; fluid CO_2 Pressure programme 190 bar isobaric for 10 min, then programmed from 190 to 290 bar at 5 bar min⁻¹.Compounds: trilaurin, trimyristin, tripalmitin, tristearin. (From J. High Resolut. Chromatogr., Chromatogr. Commun., 9 (1986) 189, with permission.)

later work, the same group used [90] similar columns but
with mass spectrometric detection. This allowed the
broadened single carbon-number peaks to be resolved into
components from constituents with different degrees of
unsaturation (Figure 13).

Supercritical Fluid Chromatography Detection

For CO_2 as mobile phase, the universal detection
capability of the FID makes it the detector of choice in
the SFC of lipids and related materials. If a modifier is
used, UV detection usually at 205 nm may be employed, but
the different molar absorptivities of the analytes makes
prior calibration necessary. Other universal detectors
compatible with modified mobile phases include the light
scattering detector (LSD).

In the LSD the effluent from the restrictor of the
end of the SFC column is nebulized and the resulting fine
spray is evaporated as it passes down a heated drift tube
to leave particles of the analyte from which light, from
a continuous light source or a laser, is scattered and
detected [91]. The LSD has been successfully applied [76]
in the packed column separation of tocopherol and
glycerides both with CO_2 alone, and with methanol-modified
CO_2 as mobile phase.

The combination of SFC with MS detection is a
powerful tool for both qualitative and quantitative
analysis. A range of effective interfaces has been
developed [69]; the most simple is that in which the
restrictor is led directly into the ion source. The
greatest sensitivity is achieved if the MS detection is
in chemical ionization (CI) mode; electron impact (EI)
spectra can be obtained, but detection limits are
generally less good [92].

Cholesterol is often used [93] to test SFC-MS systems
since dehydration across the 3-4 double bond readily
occurs on contact with 'active' sites in the apparatus.
Kallio *et al.* used [90] EI SFC-MS in the analysis of
triglycerides from butter. In single-ion monitoring mode,
the molecular ions from components of the peaks
corresponding to triglycerides containing zero to three
double bonds were recorded (Figure 17). CI MS spectra of
ecdysteroids were obtained after SFC separation; to
obtain more diagnostic fragment ions heating of the mass
spectrometer jet was necessary [94].

The coupling of SFC with FTIR detection has many
potential applications in the separation and positive

identification of lipids. Two approaches are possible; a high pressure flowcell interface and a solvent elimination interface. Flowcells which allow universal and selective detection have been constructed for use with packed columns with volumes in the region of 8 μl. The SFC separation of glycerides was monitored by Perrin and Prevot [76] with a flow cell and FTIR detection at 1743 cm^{-1}. Of major concern with this interface is the IR opacity of the mobile phase. At low pressures the CO_2 absorption bands obscure [68] the IR spectrum at 3500-3800 and 2200-2500 cm^{-1}. Although few organic compounds absorb in these regions, the Fermi resonance bands, which increase in intensity, at 2070, 1944, 1387 and 1282 cm^{-1} reduce the quality of analyte IR spectra recorded for identification purposes. This problem is compounded when a modifier is added to the CO_2.

For capillary SFC the internal dimensions of the flowcell must be reduced: strictly, a volume of about 100 nl is required to avoid loss of chromatographic resolution but construction limitations and the requirements of IR sensitivity have led to a minimum volume [70] of 800 nl. The use of this cell with a make-up fluid accessory that reduces loss of chromatographic efficiency by halving the volume has been demonstrated. Capillary SFC with flow cell FTIR has been demonstrated in steroid analysis [95].

The second FTIR detection procedure involves [96] elimination of mobile phase (either pure or modified CO_2) by rapid depressurization. The non-volatile analytes separated on the capillary column are deposited as spots directly on an IR transparent support. The support is moved beneath the restrictor outlet so that chromatographed peaks are separated spatially for FTIR analysis using the beam condensing optics of a microscope accessory. This second method offers advantages in that it tolerates mobile phase additives and increases sensitivity because many spectra can be accumulated. This procedure was employed [96] in the analysis of the triglycerides of vegetable oil.

A Comparison of Supercritical Fluid Chromatography With Other Chromatographic Techniques In Lipid Analysis

Capillary GC should be employed wherever possible in an analysis, where the high diffusion rates in gases lead to the greatest resolution and speed. The applicability of GC in lipid analysis is somewhat limited, however, because of limited thermostability of many compounds. Thus, long-chain fatty acids are converted to methyl esters before GC analysis and steroids with often multiple hydroxyl substituents are also derivatized. SFC, which offers high resolution separations without derivatization, is clearly advantageous in the analysis of such compounds. For very high molecular weight lipids [84] such as phospholipids or glycosphingolipids with molecular weight up to 3000, SFC after methylation is the only high resolution method available.

For triglycerides, the position is less clear-cut. While various authors have claimed [97] that, particularly unsaturated triglycerides can degrade [97] or polymerize [87] at the elevated temperatures necessary for elution in GC (350-400°C) others have shown routine application of high temperatures GC in direct triglyceride analysis. Sandra *et al.* [87] have pointed out that since a GC column can be operated much nearer its optimal linear mobile phase velocity than can an SFC column, very short analysis times with better resolution are possible in GC compared with SFC; Figure 14,for example compares the analysis of the triglycerides of milk chocolate by the two methods. Nonetheless, quantitation in the GC analyses of oils containing highly unsaturated triglycerides is likely to be more difficult, and it is probable that, in any case, repeated operation of capillary GC columns at temperatures up to 350°C results in impaired performance even if peak areas and retention times show small deviations [86] over consecutive runs.

Excellent selectivity can be achieved in HPLC analysis of lipids, e.g. triglycerides can be well resolved on silver ion loaded columns [98]. However, the absence of a strong chromophore in many lipid molecules means that sensitivity is low during HPLC analysis with UV detection, while response is not uniform. Refractive index detection can be used, but not for sensitive detection nor with gradient elution, and the LSD detection is gaining popularity [99].

In Figure 15 the separation [100] of a number of anabolic steroids by HPLC, and by SFC on both packed and capillary columns is compared. The most rapid analysis was achieved by packed column SFC, although with incomplete resolution; modification of the CO_2 with methanol was necessary to overcome the activity of column packing, and therefore detection was by UV absorption. The capillary column SFC separation took longer than that by HPLC, but resolution was complete; since unmodified CO_2 was the mobile phase, detection by FID was possible. Separation of steroid isomers by capillary SFC on a liquid crystalline stationary phase has been shown to be related to molecular shape [101].

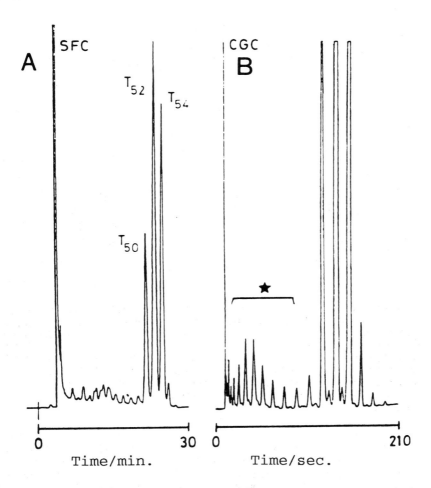

Figure 14. Analysis of triglycerides of milk chocolate by (A) capillary SFC (conditions as in Figure 16) and (B) capillary GC (column: 6 m x 25 i.d. OV-1. Temperature programme 290 to 350°C at 20°C min^{-1}; carrier gas H$_2$). (From Ref. [87], with permission.)

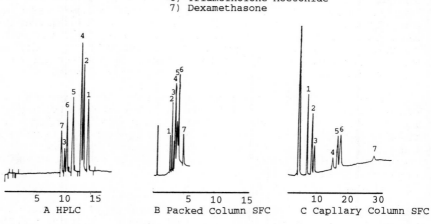

1) Melengestrol Acetate
2) Medroxyprogesterone
3) Trenbolone
4) Diethylstilbestrol
5) Zeranol
6) Triamcinolone Acetonide
7) Dexamethasone

Figure 15. Analysis of a mixture of anabolic steroids by: (A) HPLC (column: octylsilane; mobile phase methanol/acetonitrile containing 20 nM ammonium formate; detection UV at 254 nm); (B) packed column SFC (column: cyanopropyl; mobile phase methanol modified CO_2 at 60°C; detection UV at 220 nm); (C) capillary DGV (column: 5 m x 50 μm cyanoprylmethylsiloxane; capillary mobile phase pressure programme CO_2 at 70°C; detection FID). (From Ref. [100], with permission.)

Particular advantages of capillary SFC over HPLC in this context are thus the applicability of FID detection for ready quantitation, and the much simpler coupling to both FTIR and MS for compound identification. Application of SFC to a variety of lipids from wide-ranging origins [102-104] seems likely.

11. PREPARATIVE SUPERCRITICAL FLUID CHROMATOGRAPHY

If small quantities (milligrams) are required (for structure determination etc.), preparative scale SFC may be carried out [105] using conventional analytical packed column equipment. Separated materials may be collected, if non-destructive detection is employed, in vessels in which depressurization of the column effluent occurs; a simple method is to immerse the restrictor tip in a suitable solvent and recover the material of interest by evaporation.

For industrial scale production (1 g to 1 kg per hour) larger scale equipment is necessary. Extensive discussions have appeared [105-108] on the technology required to operate columns with diameters up to 100mm and packed by axial compression; commercial systems are on the market [107, 109]. Any mobile phase used in analytical packed column SFC can be used in preparative SFC, but most work so far reported has involved CO_2, chlorofluorocarbons, light hydrocarbons and CO_2 modified with alcohols, etc. The mobile phase is separated from the solute at the end of the column and recycled.

Particular attention has been paid to the means of collecting separated fractions so that products may be recovered free from solvent. After depressurization, the precipitated product may be separated [106] from the gaseous eluent by means of gravity or by means of a cyclone. Alternatively, the eluent may be collected while supercritical, frozen and the CO_2 allowed to sublime to leave the product.

Preparative SFC has been employed in the separation of a number of lipids and related compounds [105] including fatty acids [110]. Of particular interest is the purification [106, 111] of polyunsaturated fatty acid (PUFA) esters for medical and dietetic study in the prevention of heart disease. Fish oil methyl ester concentrates containing eicosapentaenoic acid (EPA) and docosahexaenoic acid (DHA) as methyl esters were separated on a 6 cm x 60 cm column packed with ODS modified silica with CO_2 mobile phase. Fractions containing EPA and DHA at, respectively, 56 and 78% purity were obtained at rates of 7.7 and 3.8 g h^{-1}.

Smaller dimension columns may be employed in preparative SFC if the eluent is recycled [107]. α- and β-tocopherols at, respectively, 85 and 70% purity were separated from wheat germ oil (itself obtained by supercritical fluid extraction) by recycle SFC on 10mm silica gel columns [112].

REFERENCES

1. Brunner, E., Hultenschmidt. G. and Schlichtharle, *J. Chem. Thermodynam.*, <u>19</u> (1987) 273.
2. Kurnik, R. T. and Reid, R. C. *Fluid Phase Equil.*, <u>8</u> (1982) 93.
3. Bartle, K. D., Clifford. A. A., Hawthorne, S. B., Langenfield, J. L., Miller, D. J. and Robinson, R. *J. Supercrit. Fluids*, <u>3</u> (1990).
4. Tyrell, H. J. V. and Harris, K. R. *Diffusion in Liquids*. Butterworths, London, 1984.

5. Reid, R. C., Prausnitz, J. M. and Sherwood, T. K. The Properties of Gases and Liquids. McGraw-Hill, New York, 1977.
6. Hawthorne, S. B. and Miller, D. J. Anal. Chem., 59 (1937) 1705.
7. Hawthorne, S. B., Miller, D. J., Walker, D and Wittington, D. in preparation.
8. McHugh, M. A., Krukonis, V. J. Supercritical Fluid Extraction-Principles and Practice. Butterworth, Stoneham, 1986.
9. Bartle, K D., Clifford, A. A., Jafar, S. A. and Shilstone, G. F. J. Phys. Chem. Ref. Data, to be published.
10 Johnston, K. P. and Eckert, C. A., AICHEJ., 27 (1981) 773.
11 Van Leer, R. A. and Paulaitis, M. E. J. Chem. Eng. Data, 25 (1980) 257.
12 Kurnik, R. T., Holla, S. J. and Reid, R. C. J. Chem. Eng. Data, 26 (1981) 47.
13 Ohgaki, K., Tsukahara, 1., Semba, K. and Katayama, T. Kagaku Kogaku Ronbunshu, 13 (1987) 298.
14 Tolley, K. K. and Tester, L. S. US Bureau of Mines Report of Investigations, 9216 (1989).
15 Moradinia, 1. and Teja, A. S. Fluid Phase Equil., 28 (1986) 199.
16 Kosal, E. and Holder, G. D., J. Chem. Eng. Data, 32 (1987) 148.
17 Stahl, E., Schilz, W., Schutz, E. and Willing, F. Angew. Chem. Int. Ed. Engl., 17 (1978) 731.
18 Pritchard, A. M., Peakall, K. A., Smart, E. F. and Bignold, G. J. Water Chem. Nucl. React. Syst., 4 (1986) 233.
19 Kumar, S. K., Chhabrai, S. P., Reid, R. C. and Suter, U. W. Macromolecules, 20 (1987) 2550.
20 Dobbs, J. M., Wong, J. M. and Johnston, K. P. J. Chem. Eng. Data, 31 (1986) 303.
21 McHugh, M. A., Seckner, A. J. and Yogan, T. J. Ind. Eng. Chem. Fundam., 23 (1984) 493.
22 Schafer, K. and Baumann, W., Fresenius B. Anal. Chem., 332 (1988) 122.
23 Sako, S., Shibata, K., Ohgaki, K. and Katayama, T. J. Supercrit. Fluids, 2 (1989) 3.
24 Chrastil, J. J. Phys Chem., 86 (1982) 3016.
25 Smith, R. D., Usdeth, H. R., Wright, D. W. and Yonker, C. R. Sep. Sci. Tech., 22 (1987) 1065.
26 Barker, 1. K., Bartle, K. D. and Clifford, A. A. Chem. Eng. Commun., 84 (1988) 4487.
27 Bartle, K. D., Clifford, A. A. and Jafar, S A. J. Chem. Soc., Faraday Trans., 86 (1990) 855.
28 Bartle, K. L., Clifford, A. A. and Jafar, S. A. J. Chem. Eng. Ref Data, to be published.
29 Peng, D. Y. and Robinson, D. B. Ind. Eng. Chem.

Fundam., 15 (1976) 59.

30   Johnston, K. P., Peck, D. G. and Kim, S. Ind. Eng.
     Chem. Res., 28 (1989) 1115.
31   Bartle, K. D., Clifford, A. A. and Shilstone, G. F.
     J. Supercrit. Fluids, 2 (1989) 30.
32   CRC Handbook of Chemistry and Physics, 52nd Edn.
     CRC Press, Florida, 1972.
33   Bartle, K. D., Clifford, A. A. and Shilstone, G. F.
     To be presented at the 2nd International
     Conference on Supercritical Fluids, Boston, 1991.
34   Paulaitis, M. E., Penninger, J. M. L., Gray, R. D.
     and Davidson, P. Chemical Engineering
     at Supercritical Fluid Conditions. Ann Arbour
     Science, Ann Arbour, 1983.
35   Charpentier, B. A. and Sevenants, M. R. (eds),
     Supercritical Fluid Extraction and Chro-
     matography (ACS Symposium Series 366). ACS,
     Washington, DC, 1983.
36   Johnston, K. P. and Penninger, J. M. L.
     Supercritical Fluid Science and Technology (ACS
     Symposium Series 406). ACS, Washington. DC, 1989.
37   Moyler, D. A. In Distilled Beverage Flavours-Recent
     Developments. Piggott, J. R. and
     Patterson, A. (eds), Ellis Horwood, Chichester,
     1989.
38   Passino, H. Ind. Eng. Chem., 41 (1949) 280.
39   Dickinson, N. and Meyers, J. J. Am. Oil Chem. Soc.,
     29 (1959) 235.
40   Brunner, G. Fette, Seifen, Anstrichm., 88 (1986)
     464.
41   Bamberger, T., Erickson, J. C. and Cooney, C. L. J.
     Chem. Eng. Data, 33 (1988) 327.
42   Klein, T. and Schulz, Ind. Eng. Chem. Res., 28
     (1989) 1073.
43   King, J. W., Bissler, R. L. and Friedrich, J. P. In
     Charpentier, B. A. and Sevenants, M. R. (eds),
     Supercritical Fluid Extraction and Chromatography
     (ACS Symposium Series 366). ACS, Washington, DC
     1988.
44   Polak, J. T., Balaban, M., Peplow, and Phlips, A.
     J. in Johnston, W. P. and Penninger, J. M. L.,
     Carpenter, B. A. and Sevenants, M. R. (eds),
     Supercritical Fluid Science and Technology (ACS
     Symposium Series 106). ACS, Washington, DC, 1989.
45   Stahl, B., Schutz, E. and Nangold, N. K. J. Agric.
     Food Chem., 28 (1980) 1153.
46   Stahl, E. and Quirin, K. W. presented at the GVC
     Meeting, Munster, 1982.
47   Stahl, E., Quirin, K. W., Glatz, A., Gerard, D. and
     Rau, G. Ber. Bunsenges. Phys. Chem.,
     88 (1984) 900.

48   Friedrich, J. P. and Pryde, E. H. J. Am. Oil Chem.
      Soc., 61 (1984) 223.
49   Brogle, W., Chem. Ind. 385 (1982).
50   Coenen, H. and Kriegel, E. Chem. Ing. Tech., MS
      1162 (1983).
51   Tiegs, C. and Peter, S. Fette, Seifen, Anstrichm.,
      87 (1985) 231.
52   Brunetti, L., Daghetta, A., Fedeli, E., Kikic, 1.
      and Zanderighi, L. J. Am. Oil Chem. Soc.,
      66 (1989) 209.
53   Gmur, W., Bosset, J. O. and Plattner, E.
      Lebensm.-Wiss u.-Technol., 19 (1986)
      419.
54   Stahl, E. and Glatz, A. Fette, Seifen, Anstrichm.,
      86 (1984) 346.
55   Willson, W. B., Stout, V. F., Gauglitz, E. J.,
      Teeny, F. M. and Mudson, J. K. In Johnston, K. P.
      and Penninger, J. M. L. (eds). Supercritical Fluid
      Science and Technology (ACS Symposium Series 406).
      ACS, Washington, DC 1989.
56   Eisenbach, W. Ber. Bunsenges, Phys. Chem., 88 (1984)
      382.
57   Rizvi, S. S. M., Chao, R. R. and Liaw, Y. J. In
      Charpentier, B. A. and Sevenants, M. R. (eds),
      Supercritical Fluid Extraction and Chromatography
      (ACS Symposium Series 366). ACS, Washington, DC
      1988.
58   Peter, S., Schneider, M., Weidner, E. and
      Ziegelitz, R. Chem. Eng. Technol., 10 (1987) .37.
59   Hawthorne, S. B. Anal. Chem., 62 (1990) 633A.
60   King, J. W., Johnson, J. H. and Friedrich, J. P. J.
      Agric. Food Chem., 37 (1989) 951.
61   Gmur, W., Bosset, J. O. and Plattner, E. J.
      Chromatogr. 388 (1987) 335.
62   King, J. W. J. Chromatogr. Sci., 27 (1989) 355
63   Mam, K. S., Kapila, S., Pieczonka, G., Clevenger,
      T. E., Yanders, A. F., Viswanath, D. A. and Mallu,
      B. Proceedings of the International Symposium on
      Supercritical Fluids, French Chemical Society,
      Paris. France, 1988.
64   Smith, R. M. (ed.) Supercritical Fluid
      Chromatography. Royal Society of Chemistry, London,
      1988.
65   Lee, M. L. and Markides, K. E. (eds), Analytical
      Supercritical Fluid Chromatography. Chromatography
      Conferences Inc. Provo, Utah, 1990.
66   Richter, B. E., Bornhop, D. J., Swanson, J.
      T.,Wangsgaard, J. G. and Andersen, M. R. J.
      Chromatogr. Sci., 27 (1989) 303.
67   Bornhop, D. J. and Wangsgaard, J. D. J. Chromatogr.
      Sci., 27 (1989) 293.

68 Bartle, K. D., Raynor, M. W., Clifford, A. A.,
    Davies, 1. L., Kithinji, J. P., Shilstone, G. F.,
    Chalmers, J. M. and Cook, B. W. J. Chromatogr.
    Sci., 27 (1989) 283.
69 Games, D. E., Berry, A. J., Mylchreest, I. C.,
    Perkins, J. R. and Pleasance, S. In Smith, R. M.
    (ed.), Supercritical Fluid Chromatography. Royal
    Society of Chemistry, London, 1988, p. 159.
70 Raynor, M. W., Clifford, A. A., Bartle, K. D.,
    Reyner, C., Williams, A. and Cook, B. W. J.
    Microcolumn. Sep., I (1989) 101.
71 Raynor, M. W., Bartle, K. D., Davies, I. L.,
    Williams, A., Clifford, A. A., Chalmers, J. M. and
    Cook, B. W. Anal. Chem., 60 (1988) 427.
72 Smith, 1. D., Chapman, E. G. and Wright, B. W.
    Anal. Chem., 57 91985) 2829.
73 Hammond, E. W In Hamilton, R. J. and Patterson. A.
    (eds), Analysis of Fats and Oils. Elsevier, London,
    1986, p. 113.
74 Ackman, R. G. In Hamilton, R. J. and Patterson, A.
    (eds), Analysis of Fats and Oils. Elsevier, London,
    1986, p. 137.
75 Rawdon, M. G. and Norris, T. A. Intern. Laboratory,
    p. 12 (1984).
76 Perrin, J. L. and Prevot, A. Rev. Fr. Corps. Gras,
    35 (1988) 485.
77 Hellgeth, J. W., Jordan, J. W., Taylor, L. T. and
    Khorassani, M. A. J. Chromatogr. Sci . 24 (1986)
    183.
78 Liu, Y. and Yang, F. J. J. Microcolumn Sep., 2
    (1990) 245.
79 Thiebaut, D., Caude, M. and Rosset, R. Analysis, 15
    (1985) 528.
80 Bartle, K. D., Clifford, A A. and Jeffrey, B.
    unpublished measurements (1989).
81 Chester, T. L. J. Chromatogr., 299 (1984) 424.
82 White, C. M. and Houck, R. K. J. High Res.
    Chromatogr., 9 (1986) 424.
83 Raynor. M. W., Kithinji, J. P., Barker. 1. K.,
    Bartle. K. D. and Wilson, I. D. J. Chromatogr., 436
    (1988) 497.
84 Kuei, J., Her, G. R. and Reinhold, V. N., Anal.
    Biochem., 172 (1988) 228.
85 David, P. A. and Novotny, M. J. Chromatogr., 461
    (1989) 111.
86 Proot, M., Sandra, P. and Geeraert, E. J. High Res.
    Chromatogr., 9 (1986) 189.
87 Sandra, P. in Smith, R. M. (ed.) Supercritical
    Fluid Chromatography. Royal Society of Chemistry,
    London, 1988, p. 137.
88 Richter, B. E., Anderson, M. R., Knowles, D. E.,
    Campbell, E. R.. Porter, N. L., Nixon, L. and

Later, D. W. In Charpentier, B. A. and Sevenants, M. R. (eds), Supercritical Fluid Extraction and Chromatography (ACS Symposium Series 366). ACS, Washington, DC, 1988, p. 179.

89   Huopalahti, R.. Laakso, P., Saaristo, J., Linko, R. R. and Kallio, H. J. High Res. Chromatogr., 11 (1988) 899.

90   Kallio, H., Laakso, P., Huopalahti, R. and Linko, R. R. Anal. Chem., 61 (1989) 698.

91   Hoftmann, S. and Greibrokk, T. J. Microcolumn Sep., 1, 35 (1989).

92   Voorhees, K. J., Zaugg, S. D. and DeLuca, S. J. In White, C. M. (ed.), Modern Supercritical Fluid Chromatography. Huthig, Heidelberg, 1988, p. 59.

93   Owens, G. D., Burkes, L. J., Pinkston, J. D., Keough, T., Simms, J. R. and Lacey, M. P. In Charpentier, B. A. and Sevenants, M. R. (eds), Supercritical Fluid Extraction and Chromatography (ACS Symposium Series 366). ACS, Washington, DC, 1988), p.191.

94   Raynor, M. W., Kithinji, J.P., Bartle, K. D., Games, D.E. Mylchreest, I.C. Lafont, R., Morgan, E.D. and Wilson, I.D. J. Chromatogr. 467 (1989) 292.

95   Shah, S., Ashraf-Khorassani, M. and Taylor, L. T., Chromatographia, 25 (1988) 631.

96   Pentoney, S. L., Shafer, K. H., Griffiths, P. R. and Fuoco, R. J. High Res. Chromatogr.. 9 (1986) 169.

97   Mares, P., Skorepa, J., Sinkelkove, E. and Turzlcka, E. J. Chromatogr., 273 (1983) 172.

98   Christie, W. W. J. Chromatogr., 454 (1988) 273.

99   Palmer, A. J. and Palmer, F. J. J. Chromatogr., 465 (1989) 369.

100   Richter, B. E. and Knowles, D. E. Lee Scientific Applications Note Number 008. Salt Lake City, Utah, 1990.

101   Chang, H. C., Markides, K. E., Bradshaw, J. S. and Lee, M. L., J. Microcolumn Sep., 1 (1989) 131.

102   Holzer, G. C., Kelly, P. J. and Jones, W. J. J. Microbiol. Methods, 8, (1988) 161.

103   Sakaki, K., Sako, T., Yokochi, T., Suzuki, O. and Hakuta, T. Yukagaku, 37 (1988) 54.

104   Normura, A., Yamada, J., Tsunoda, K., Sakaki, K. and Yokuchi, T. Anal. Chem., 61 (1989) 2076.

105   Berger, C. and Perrut, M. J. Chromatogr., 505 (1990) 37.

106   Berger, C., Justforgues, P. and Perrut, M. Proceedings of the International Symposium on Supercritical Fluids, French Chemical Society, Paris, 1988, p. 397.

107   Saito, M., Yamauchi, Y., Hondo, T. and Senda, M.
      <u>Proceedings of the International
      Symposium on Supercritical Fluids</u>, French Chemical
      Society, Paris 1988, p. 381.
108   Alkio, M., Harvala, T. and Komppa, V. <u>Proceedings
      of the International Conference on Supercritical
      Fluids</u>, French Chemical Society, Paris, 1988, p.
      389.
109   <u>Prochrom Preparative Scale Supercritical Fluid
      Chromatography</u>, Prochrom, Champigneulles, France.
110   Jusforgues, P. and Perrut, M. European Patent
      Application EP254610 A1 (1988).
111   Alkio, M. and Komppa, V. <u>Kem.-Kemi</u>, 17 (1990) 354.
112   Saito, M. and Yamauchi, Y. <u>J. Chromatogr.</u>, <u>505</u>
      (1990) 257.

Analysis of Triacylglycerols by Argentation Supercritical Fluid Chromatography

L. G. Blomberg and M. Demibuker

DEPARTMENT OF ANALYTICAL CHEMISTRY, STOCKHOLM UNIVERSITY, ARRHENIUS LABORATORY FOR THE NATURAL SCIENCES, S-106 91 STOCKHOLM, SWEDEN

1 INTRODUCTION

Supercritical Fluid Chromatography (SFC).

Use of supercritical media as a mobile phase in chromatography was first reported by Klesper and coworkers in 1962 [1]. At that time, packed columns were employed. However, for a long period of time, the technique was studied and advanced only by a small group of scientists. It was not until 1981 with the advent of open tubular column SFC, as proposed by Novotny *et al.* [2,3], that the technique received more widespread interest. During recent years, however, techniques for packed column SFC have been greatly improved. Columns are thus packed with high quality material, originally intended for HPLC. Further, a multitude of different mobile phase programming techniques has been developed. Again, Klesper and coworkers have been the driving force in this evolution [4-6]. In addition, improved detection systems, e.g. miniaturized light scattering detection [7] and mass spectrometry [8,9] has made quantitative analysis feasible also when carbon dioxide with a modifier is applied as the mobile phase. Moreover, companies such as Hewlett Packard, Gilson Medical Electronics [10] and Jeol [9] promote the development of packed column SFC.

The utility of open tubular SFC has been demonstrated in a number of applications [11,12]; however, the low diffusion in supercritical media at high densities currently causes a restriction of this technique. Low diffusion leads to low optimal flow rates and high C_m-terms of the Golay equation. This results in long analysis times or, alternatively, the separation efficiency has to be sacrificed. This effect can be compensated for by the application of smaller capillary diameters, and thus 50 μm i.d. is currently being used. However, in order to be comparable with the state of the art in GC, column diameters would have to be in the range of 20 μm. For practical reasons, such diameters are not applicable. It should be mentioned that a lower i.d. leads to decreased HETP and thus, separation can be obtained on shorter column lengths, and some analysis time can be saved in this manner. In conclusion, it presently seems that packed column SFC offers a greater potential than open tubular SFC.

Packing material for HPLC is now available in quite small particle sizes, and this makes the combination of high speed and relatively high separation efficiency feasible. The attainable plate numbers are, however, limited by the pressure drop, but there are a large number of selectively-separating packing materials that may help, in many cases, to give the desired separation. Silica-based packing materials generally show some adsorption of the analytes, with the exception of pure hydrocarbons. A polar modifier has to be added to the mobile phase to avoid such adsorption. Addition of modifiers precludes the application of the flame ionization detector, but, as mentioned above, other detectors are available for this situation.

During recent years, we have been developing a SFC method for the analysis of triacylglycerols (TG). Initially, separation was attempted on micro-packed, i.d. 0.25 mm, silver-ion impregnated columns, UV was employed for the detection, and density programming was by negative temperature and positive pressure programming [13-17]. The next step was the construction of a miniaturized light scattering detector in order to facilitate quantitative analysis [7]. Column i.d. was then 0.7 mm. The type of mobile phase programming that we had employed was not by any means optimal for silver ion chromatography. First, negative temperature programming leads to the formation of stronger complexes between the unsaturated solutes and the stationary phase as the program proceeds, thus counteracting the very purpose of the gradient. Second, as the density of the mobile phase increases, the proportion of modifier content decreases, which also contradicts the intention with the program. Application of a modifier gradient will be more beneficial [18].

2 EXPERIMENTAL

Columns. Fused silica capillaries, i.d. 0.25 mm (Polymicro Technologies, Phoenix, AZ), or glass-lined metal tubing, i.d. 0.7 mm (SGE, Ringwood, Victoria, Australia) were used. The columns were slurry-packed with Nucleosil 5 SA or 4 SA (Macherey Nagel, Düren, Germany) and impregnated with silver nitrate as described earlier [13,14]. It should be noted that the packed bed is compacted on the treatment with the aqueous solution. The practical aspects of such a compaction were discussed recently [19]. Fused silica capillary tubing (Polymicro Technologies), 9 or 10 μm i.d. was used as a restrictor.

<u>Chromatography.</u> Chromatography was performed on a Lee Scientific 600 Series SFC (Salt Lake City, UT). The mobile phase consisted of a mixture of carbon dioxide, acetonitrile, isopropanol, 92.8:6.5:0.7 mol %, for isocratic elution of triacylglycerols, and 97.1:2.6:0.3 mol % for fatty acid methyl esters (FAME). SFC grade carbon dioxide (Scott Spec. Gases, Plumsteadville, PA) was used; however, a less expensive grade, 99.99 %, was also suitable for use. Negative temperature and positive pressure programming as well as compositional gradients were applied. For eluent compositional gradients, a second pump, Schimadzu, LC-10AD, was via a mixing chamber, TCMA0120113T (The Lee Company, Westbrook, CT), connected to the SFC, Figure 1. The second pump was run in a constant flow mode, while the first pump had to be run in a constant pressure mode (this pump could not be operated in the constant flow mode). The split ratio was 1:4 and the mobile phase velocity was *ca* 5 mm/s. Isocratic elution was performed at *ca* 3.5 mm/s.

<u>Detection.</u> Separation was followed by means of detection on a Lee Scientific UV detector at 210 nm or by a miniaturized light scattering detector [7]; the construction of the latter detector is shown in Figure 2. The detector cell was made of black PVC-plastic, and the diameter of the cell compartment was 30 mm.

<u>System Evaluation.</u> Standard substances and chromatographically purified oils were obtained from Larodan Fine Chem. (Malmö, Sweden), cohune oil from LipidTeknik (Stockholm, Sweden) and hydrogenated oils from Karlshamns Oils and Fats AB (Karlshamn, Sweden). The analytes were dissolved in HPLC-grade pentane in concentrations of 30 mg/mL. When necessary, addition of chloroform was made. For UV-detection and with columns having an i.d. of 0.25 mm, injection was made with a split ratio of 1:1 and a timed split of 0.2 s. Using these conditions, ca 60 nL was allowed to enter the column. Sample injection was 200 nL without split when columns with an i.d. of 0.7 mm and ELSD were applied. Peak identifications were tentative, and were based on the retention of standards and the separation of fractions obtained with reversed phase HPLC [15]. Further, the retention times of different TG were established by comparison of the chromatograms from a number of different vegetable oils. The composition of these oils had first been determined by means of reversed phase HPLC. The acyl groups are, in this work, given in an arbitrary sequence, and do not denote any specific position at the glycerol moiety.

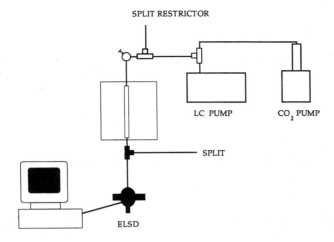

Figure 1. System for eluent composition programming.

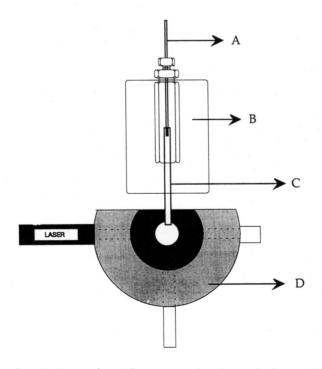

Figure 2. Schematic diagram of the miniaturized light scattering detector. A = restrictor; B = heating block; C = drift tube; D = detector cell.

3 RESULTS AND DISCUSSION

Mobile Phase Considerations.

It was necessary to add acetonitrile to the carbon dioxide to facilitate elution. Acetonitrile modifies the retentive properties of the stationary phase and improves the solubility of analytes in the mobile phase. A minor amount of isopropanol was added in order to achieve a mobile phase that is homogeneous under the applied conditions. It should be pointed out that reproducible retention times can be obtained only when the mobile phase consists of one single phase.

Addition of modifiers to the carbon dioxide generally results in an increase in critical temperature (T_c) and pressure (P_c). Calculation of the critical parameters for the mixture used for separation of triacylglycerols under isocratic conditions gave T_c = 62°C and P_c = 101 atm [14]. When operating under conditions below critical, the technique is not SFC in *sensu strictu*; however, also liquid carbon dioxide provides relatively high diffusion and the viscosity is lower than for liquids ordinarily used as mobile phases for HPLC. Thus, also chromatography using sub-critical carbon dioxide as mobile phase may provide faster and more efficient separations than conventional HPLC.

Retention in Argentation SFC.

Retention characteristics in argentation TLC and HPLC have been extensively discussed [20-23]. Clearly, the retention can be influenced by the nature of the mobile phase, and thus some mobile phases can reduce the double bond effect. In SFC, a dienoyl residue has a lower retention value than two monoenes in the same molecule and a triene than a monoene and a diene, e.g. SOL is eluted before OOO, Figure 3. Christie reported the reversed elution order for argentation HPLC [24,25]. Some degree of chain length separation is sometimes observed in argentation HPLC [22]; in SFC this type of separation can be extensive, Figure 4. Separation of triacylglycerols differing only in the position of one double bond in one fatty acid moiety was demonstrated recently [17]. Separation of TG differing from each other only in the position of a double bond has been achieved also by argentation HPLC [26,27].

Partial hydrogenation of an oil will result in a great number of *trans* isomers; group separation of partially hydrogenated oils is shown in Figure 5.

Composition Gradients.

An advantage of SFC is that a great number of gradients affecting the properties of the mobile phase

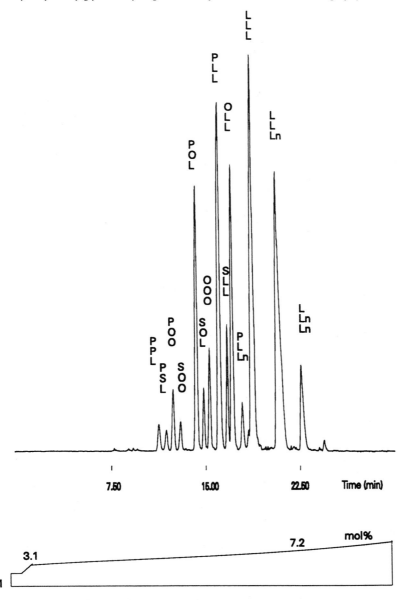

Figure 3. Supercritical fluid chromatogram, evaporative light scattering detection, of soybean oil. Column: glass-lined metal tubing, 150 mm x 0.7 mm, packed with Nucleosil 4 SA and impregnated with silver nitrate. Conditions: temperature 100°C; pressure 340 atm.; mobile phase: gradient of carbon dioxide / (acetonitrile-isopropanol, 90:10 mol %); restrictor: fused silica capillary tubing 130 mm x 10 μm. P = palmitate; S = stearate; O = oleate; L = linoleate; Ln = α-linolenate.

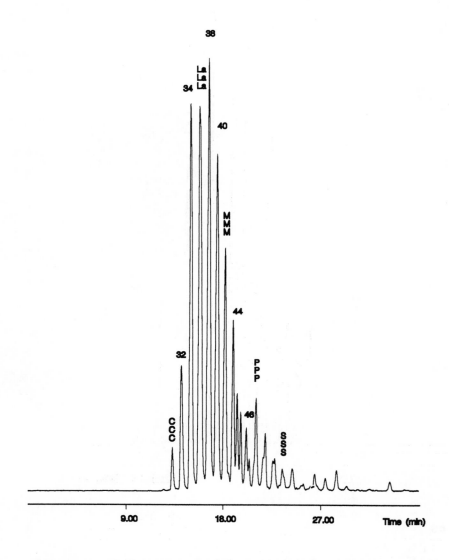

Figure 4. Supercritical fluid chromatogram, evaporative light scattering detection, of cohune oil. Column as in Figure 3. Conditions: injection at 115°C and 180 atm, after 1 min, programmed at -1°C/min to 75°C and at 2 atm/min to 210 atm, and then 4 atm/min to 320 atm. Mobile phase: carbon dioxide-acetonitrile-isopropanol (92.8:6.5:0.7) mol %. Peaks: CCC = tricaprin; LaLaLa = trilaurin; MMM = trimyristin; PPP = tripalmitin; SSS = tristearin. Peak numbers refer to carbon numbers.

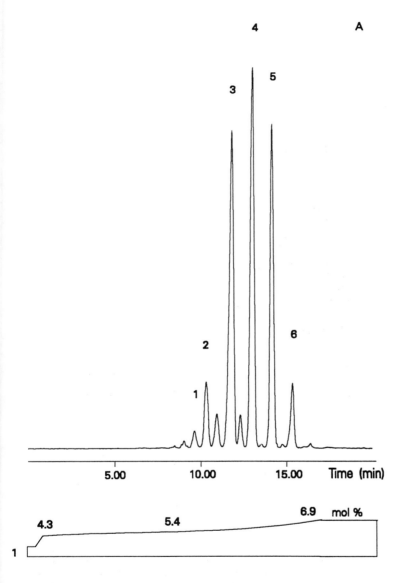

Figure 5. Supercritical fluid chromatograms, evaporative light scattering detection, of partially hydrogenated oils, A = rapeseed oil (Lobra); B = a fish oil. Column and conditions as in Figure 3. Peaks: 1 = P(two *trans*-monoenes); 2 = S(two *trans*-monoenes); 3 = three *trans*-monoenes; 4 = one *cis*-, two *trans*-monoenes; 5 = two *cis*-, one *trans*-monoene; 6 = three *cis*-monoenes. Reproduced from Ref. 18 with permission.

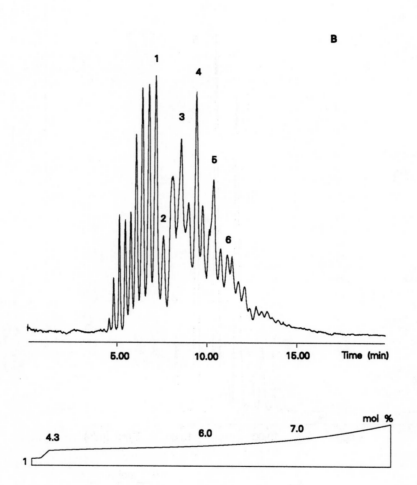

Figure 5. Supercritical fluid chromatograms, evaporative light scattering detection, of partially hydrogenated oils, A = rapeseed oil (Lobra); B = a fish oil. Column and conditions as in Figure 3. Peaks: 1 = P(two *trans*-monoenes); 2 = S(two *trans*-monoenes); 3 = three *trans*-monoenes; 4 = one *cis*-, two *trans*-monoenes; 5 = two *cis*-, one *trans*-monoene; 6 = three *cis*-monoenes. Reproduced from Ref. 18 with permission.

are applicable. Most commonly, gradients resulting in increased mobile phase densities are applied. Such gradients, however, result in a decrease in solute diffusion coefficients, which leads to impaired chromatographic performance. The optimal mobile phase velocity, u opt, will thus be decreased and the slope of the high velocity branch of the van Deemter curve will have a higher gradient. Moreover, pressure programming, without application of constant flow regulation, results in increased mobile phase velocities, leading to decaying separation efficiencies. On the contrary, it would be beneficial to apply a program that would decrease the mobile phase flow rate [6]. When applying positive pressure and negative temperature gradients, we have thus been obliged to employ slow mobile phase velocities, typically 2 - 3 mm/s, at the start of a run, in order to maintain the highest separation efficiency throughout the entire analysis. In addition, negative temperature programming is not really optimal in connection with argentation chromatography, since the strength of the olefin/silver ion complex thereby will be increased.

Application of moderate modifier gradients on packed column SFC is an attractive approach. With such gradients, the mobile phase elution strength will be greatly enhanced, while diffusion coefficients will be only moderately increased. Thereby, it will be possible to apply higher mobile phase velocities without appreciable losses in separation efficiency. Further, when no constant flow regulation is employed, the velocity will be decreased during the composition program, which may be an advantage, since late-eluting compounds, in general, have a lower u opt. Moreover, it may be advantageous to combine a compositional gradient with a positive temperature gradient [5]. Such temperature gradients could be valuable in argentation chromatography. The selectivity will, however, be lost at temperatures above 125°C. The advantages of mobile phase composition programming are demonstrated in Figure 6.

The chromatographic system applied in the present work was not optimal for accurate compositional gradient elution. The accuracy of the reciprocating Schimadzu pump was thus only ± 2 μL/min. This could partly be overcome by the application of a split as in Figure 1. In order to obtain a fast response to changing mobile phase composition, the dead volume between split and column must be quite small. Further, due to the low mobile phase flow rate, at least 30 min of system equilibration was needed between runs in order to obtain retention time reproducibility. The retention time RSD of peak 5, Figure 5A was 4.35 (four runs). With the current instrumentation, the use of columns having a somewhat larger i.d. would result in improved gradient reproducibility. For narrow bore

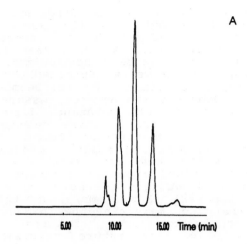

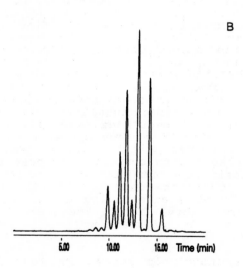

Figure   6.   Supercritical   fluid   chromatograms,
evaporative light scattering detection of hydrogenated
soybean oil. Column as in Figure 3. Conditions: A =
temp./pressure programmed; B = composition programmed.

columns, the use of a flow regulated syringe pump for the mobile phase modifier should lead to improved precision. Columns having an i.d. of 1 - 2 mm can be employed when a split is inserted between the column and the detector. Such a split could, when it is easily regulated, be opened after a programmed run in order to achieve a rapid equilibration before the following injection. Moreover, a second detector, e.g. UV or MS could be applied to the split line [9].

Application of a mobile phase gradient is beneficial for separation of oils containing components that have a widely differing degree of unsaturation. For isocratic elution, a relatively high concentration of modifier is required to elute the more highly unsaturated species. Under such conditions, species having few unsaturations elute quickly without being well-separated, *cf.* separation of fish oil described in Refs. 13 and 28. A great improvement was achieved when applying a mobile phase composition gradient, Figure 7. Further, a general advantage of eluent composition programming is that, after injection, the analytes are focused at the column head, which, of course, results in improved column performance.

Evaporative Light Scattering Detection.

Detection is often a problem with packed column SFC. Flame ionization detection (FID) can be used when neat carbon dioxide is being employed as mobile phase. However, for most samples, it is necessary to add a polar modifier to the mobile phase, and this precludes the use of the FID. Detection by means of UV is therefore often employed. For the present application, such detection is not suitable, since the UV-response is proportional to the number of double bonds and saturated TG cannot be detected at all. The use of an evaporative light scattering detector (ELSD) makes it possible to solve the detection problem.

The ELSD has originally been constructed for connection to conventional HPLC [29], but the detector has also been used with packed column SFC, using wide [30-34] or narrow bore columns [35,36,7]. The response of the detector is highly dependent on the mobile phase flow rate. Carraud *et al.* [30] thus obtained a maximum in a plot of peak area *vs* mobile phase flow rate. It may be speculated that the ascending section of the curve is caused by losses of analytes on the walls of the drift tube at low flow rates, while the descending section may be explained by the formation of smaller drops at high flow rates. It was deduced that pressure-programming in SFC could not be combined with connection to ELSD [30]. In SFC, pressure programming is of crucial importance for the optimization of the separation, and it was decided to develop an ELSD that would give a similar response over a range of mobile

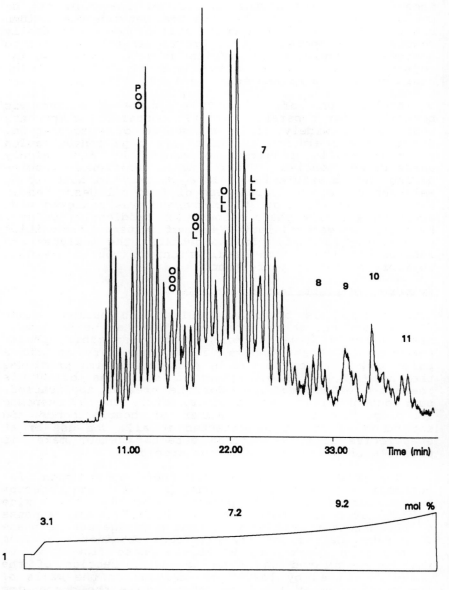

Figure 7. Supercritical fluid chromatogram, evaporative light scattering detection, of CPL fish oil 30. Column and conditions as in Figure 3. Peak numbers refer to the number of double bonds. Reproduced from Ref. 18 with permission.

phase flow rates. This would, of course, also improve detector performance when applying flow programming. The approach that was taken was to try to reduce the losses of solutes in the drift tube at low mobile phase flow rates. For this purpose, we took advantage of the low mobile phase flow that we had from the micro columns; a miniaturized drift tube would, in this case, be sufficient for the evaporation of the mobile phase. Detector dimensions were thus optimized for connection to packed columns having an i.d. of 0.7 mm. As indicated above, wider columns can also be employed, but a split should then be installed between column and detector. The application of 4.6 mm i.d. columns would allow split ratios of *ca* 1:100, thus providing a convenient method for fractionation.

The response of the miniaturized detector was relatively constant over a range of flow rates of expanded mobile phase, 8 to 16 mL/min, and also at flow rates above 18 mL/min [7]. The limit of detection for trimyristin, triolein and trilinolenin was less than 6 ng when the lower range of flow rates was applied. This is in the same range as reported for other ELSD systems. There is, however, a potential for further improvement of the LOD of our detector. At the higher flow rates, the response was, however, ca 20 times lower.

Several different types of restrictors have been employed in SFC [12]. All of these have their advantages and disadvantages. In this work, narrow bore fused silica capillaries have been used as restrictors. The main advantage of such restrictors, in this context, is that they are easy to reproduce. This enables the formation of a reproducible spray of mobile phase in the detector, which is essential for the detector performance. The main disadvantage of open tubular restrictors is that solutes, provided they are poorly soluble in the mobile phase, may, to some slight extent, be deposited in the restrictor. In some cases, minor spikes can be observed, *e.g.* in the chromatogram of cohune oil in Ref. 7. However, these can be removed by some gentle signal filtering *cf.* Figure 4.

The suitability for quantitative analysis was demonstrated for corn oil [7]. The relative standard deviation in peak areas was generally less than 4 % (n = 6). Minor components had peak areas with higher RSD. Using isocratic conditions, the coefficient of determination of a log/log plot of peak areas *vs* sample amount was 0.9940 over a range of 11 to 200 ng.

Also mobile phase compositional gradients may affect the detector response. This has been attributed to changes in droplet size due to alterations of mobile phase surface tension and viscosity [37]. Here, response factors have to be employed.

A method has been described where supercritical fluid chromatography is applied for the quantitative analysis of triacylglycerols. The decisive point is the comparison of the performance for the particular application with that obtained using other chromatographic techniques. First, the advantage over GC is that polyunsaturated TG are stable under SFC conditions but not at the high temperatures necessary for separation by means of GC. Second, SFC possesses a higher intrinsic separation power than HPLC, which we have attempted to demonstrate in this article. Separation can thus be performed more efficiently and analysis times can be shorter. It could be argued, however, that chain length separations are not always desirable, since they complicate relatively simple traces. A second factor of decisive importance is the commercial availability of the necessary equipment. It seems that the recent generation of SFC instruments, dedicated to packed column operation, in combination with commercially available ELSD would be suitable for the quantitative analysis of triacylglycerols.

For several years, HPLC has been the general method for the analysis of polyunsaturated TG, and excellent results are being obtained in reversed phase as well as argentation modes [38,39]. However, packed column argentation SFC is a technique that deserves to be considered as complementary to HPLC for the analysis of triacylglycerols.

3 ACKNOWLEDGEMENTS

Financial support from Karlshamn Research Council, Karlshamn, Sweden is gratefully acknowledged. Thanks are due to Per Andersson for technical assistance with the light scattering detector.

4 REFERENCES

1. E. Klesper, A.H. Corwin and D.A. Turner, *J. Org. Chem.*, 1962, **27**, 700.
2. M. Novotny, S.R. Springston, P.A. Peaden, J.C. Fjeldsted and M.L. Lee, *Anal. Chem.*, 1981, **53**, 407A.
3. M. Novotny, M.L. Lee, P.A. Peaden, J.C. Fjeldsted and S.R. Springston, *U.S. Pat.*, 4,479,380, (1984).
4. E. Klesper annd F.P. Schmitz, *J. Chromatogr.*, 1987, **402**, 1.
5. F.P. Schmitz and E. Klesper, *J. Chromatogr.*, 1987, **388**, 3.
6. S. Küppers, M. Grosse-Ophoff and E. Klesper, *J. Chromatogr.*, 1993, **629**, 345.
7. M. Demirbüker, P.E. Andersson and L.G. Blomberg, *J. Microcol. Sep.*, 1993, **5**, 141.
8. J.D. Pinkston, T.E. Delaney, K.L. Morand annd R.G. Cooks, *Anal. Chem.*, 1992, **64**, 1571.

9. M. Takeuchi and T. Saito, in K. Jinno (Editor), *Hyphenated Techniques in Supercritical Fluid Chromatography and Extraction*, J. Chromatogr. Library **53**, Elsevier, Amsterdam, 1992, pp. 239-271.
10. F. Vérillon, D. Heems, B. Pichon, H. Coleman and J.-C. Robert, Am. *Lab.* (Fairfield) 1992, **24**, No. 9, 45.
11. K.E. Markides and M.L. Lee, *SFC Applications, Symposium/Workshop on Supercritical Fluid Chromatography*, Provo, UT, 1989.
12. M.L. Lee and K.E. Markides (Editors) *Analytical Supercritical Fluid Chromatography and Extraction*, Chromatography Conferences Inc., Provo, UT, 1990.
13. M. Demirbüker and L.G. Blomberg, *J. Chromatogr. Sci.*, 1990, **28**, 67.
14. M. Demirbüker and L.G. Blomberg, *J. Chromatogr.*, 1991, **550**, 765.
15. M. Demirbüker, L.G. Blomberg, N.U. Olsson, M. Bergqvist, B.G. Herslöf and F. Alvarado Jacobs, *Lipids*, 1992, **27**, 436.
16. M. Demirbüker, I. Hägglund and L.G. Blomberg, *J. Chromatogr.*, 1992, **605**, 263.
17. M. Demirbüker, *Thesis*, Stockholm University, Stockholm, 1992.
18. L.G. Blomberg, M. Demirbüker and P.E. Andersson, *J. Am. Oil Chem. Soc.*, 1993, **70**, No 10.
19. P.E. Andersson, M. Demirbüker and L.G. Blomberg, *J. Chromatogr.*, 1993, **641**, 347.
20. L.J. Morris, *J. Lipid Res.*, 1966, **7**, 717.
21. L.J. Morris and B. Nichols, in A. Niederwieser (Editor) *Progress in Thin Layer Chromatography, Related Methods*, Ann Arbor Humphrey Sci., Ann Arbor, MN, (1972), pp. 74-93.
22. B. Nikolova-Damyanova, in W.W. Christie (Editor), *Advances in Lipid Methodology - One*, The Oily Press, Ayr, 1992, pp. 181-237.
23. D. Cagniant, in D. Cagniant (Editor), *Complexation Chromatography*, Marcel Dekker, New York, 1992, pp. 149-195.
24. W.W. Christie, *J. Chromatogr.*, 1988, **454**, 273.
25. W.W. Christie, *Rev. Corps Gras.*, 1991, **38**, 155.
26. B. Nikolova-Demyanova, W.W. Christie and B.G. Herslöf, *J. Am. Oil Chem. Soc.*, 1990, 67, 503.
27. W.W. Christie, *Fat Sci. Technol.*, 1991, **93**, 66.
28. M. Demirbüker, I. Hägglund and L.G. Blomberg, in N.U. Olsson and B.G. Herslöf (Editors), *Contemporary Lipid Analysis*, LipidTeknik, Stockholm, 1992, pp. 30-47.
29. W.W. Christie, in W.W. Christie (Editor), *Advances in Lipid Methodology - One*, Oily press, Ayr, 1992, pp. 239-271.
30. P. Carraud, D. Thiebaut, M. Caude, R. Rosset, M. Lafosse and M. Dreux, *J. Chromatogr. Sci.*, 1987, **25**, 395.
31. M. Lafosse, M. Dreux and L. Morin-Allory, *J. Chromatogr.*, 1987, **404**, 95.
32. D. Upnmoor and G. Brunner, *Chromatographia*, 1992,

33, 255.
33. M. Lafosse, C. Elfakir, L. Morin-Allory and M. Dreux, *J. High Resolut. Chromatogr.,* 1992, **15,** 312.
34. S. Brossard, M. Lafosse and M. Dreux, *J. Chromatogr.,* 1992, **623,** 323.
35. S. Hoffmann and T. Greibrokk, *J. Microcol. Sep.,* 1989, **1,** 35.
36. H.M. Hagen, K.E. Landmark and T. Greibrokk, *Ibid.* 1991, **3,** 27.
37. S. Hoffmann, *Thesis,* Stockholm University, Stockholm, 1989.
38. W.W. Christie, *HPLC and Lipids,* Pergamon Press, Oxford, 1987.
39. W.W. Christie (Editor) *Advances in Lipid Methodology - One,* Oily Press, Ayr, 1992.

High-performance Liquid Chromatography of Chiral Lipids

W. W. Christie

THE SCOTTISH CROP RESEARCH INSTITUTE, INVERGOWRIE,
DUNDEE DD2 5DA, UK

1. INTRODUCTION

In recent years, the chromatographic behaviour of chiral lipids or their derivatives has made a significant contribution to the solution of stereochemical problems, and made it much easier to obtain pure enantiomers for biological studies. Many factors have contributed to the possibilities for improved chromatographic resolution, including technical developments such as the design of the many components of high-performance liquid chromatography (HPLC) systems, e.g. pumps, injection valves and detectors, that have lead to great improvements in the capabilities and efficiency of this technique.[1] Most important of all has been the knowledge gained of the effects of chirality on chromatographic selection mechanisms, especially in the laboratory of W.H. Pirkle. This has lead to the discovery of suitable derivatives for chiral resolution and to the development of novel chiral stationary phases for HPLC. The topic of chromatographic resolution of chiral lipids has been reviewed recently by the author[2] and by Takagi[3], so this review deals mainly with general principles and some relatively new applications.

It is axiomatic that a molecule has the property of chirality or asymmetry when its mirror image cannot be superimposed on itself. The two non-superimposable forms are termed *enantiomers*; they have the same chemical bonds but differ in the arrangement of these in space and thus in their three-dimensional structures. As they have identical chemical properties, they interact with reagents that are non-chiral in exactly the same way, although any products formed have different molecular configurations. On the other hand, each enantiomer may react at a different rate with a given enantiomeric form of a chiral reagent. Molecules with more than one centre of asymmetry can exist in 2^n forms, where n is the number of asymmetric centres. Stereoisomers in which the configuration of one or more (but not all) of the asymmetric centres differs are termed *diastereomers*. With a given compound, diastereomers are not mirror images and may have slightly different conformations and thence different internal energies.

In order to effect the resolution into enantiomers by chromatographic means, it is therefore necessary that they be made to interact with chiral molecules in either the mobile or the stationary phase of whatever system is adopted. For example, enantiomers may interact with a chiral stationary phase to form transient diastereomeric complexes, which are potentially resolvable. Diastereomeric derivatives *per se* have distinct chemical and physical properties and thus have the potential to be separated in non-chiral chromatographic environments.

Lipids like all natural compounds are synthesised by enzyme systems which invariably produce chiral products. For example in isoprenoid and other branched-chain lipids, the asymmetry may be inserted directly from chiral precursors during the biosynthesis of the aliphatic chains. Alternatively, the chirality may be introduced to the aliphatic chain later as in the synthesis of oxygenated or cyclopropane fatty acids. The nature and the absolute configurations of the aliphatic constituents of lipids, including fatty acids, have been reviewed by Smith.[4] The glycerol moiety of glycerolipids also has a centre of asymmetry at the second carbon atom, and their stereochemistry has also been reviewed by Smith[5] (see also appendix).

The primary method for the determination of the stereochemical configuration of a chiral molecule is chemical synthesis from a precursor of defined stereochemistry by procedures that do not cause inversion of the asymmetric centres, followed by optical measurements. Usually, synthesis is a technically daunting task and comparisons of the sign of the Cotton effect in the optical rotatory dispersion curves of unknowns with similar compounds of defined configuration is a valuable alternative. Nuclear magnetic resonance spectroscopy with chiral shift reagents is being used increasingly for the same purpose. However, for most natural lipids the magnitude of the observed effect is too small to be of practical value. Chiral chromatography can now often substitute for these "classical" methods.

Pirkle and House[6] in 1979 provided a theoretical explanation of the "three-point rule" and put chiral chromatography on a systematic basis, *viz*. if chiral resolution is to be effected, there must be at least three simultaneous interactions between a solute enantiomer and a chiral stationary phase and one must be stereochemically dependent. Those intermolecular forces influencing chiral resolution include hydrogen bonds, *pi-pi*, dipole-dipole, steric and hydrophobic interactions. In effect, the solute and the stationary

phase form a short-lived diastereomeric complex, and
the configurations of the substituents in each
enantiomer determine the strengths of the interactions,
and thence the order of elution and ultimately whether
resolution is achievable. It is thus possible to tailor
the solute by appropriate derivatisation to maximise
the effect. For example, an advantageous strategy for
effecting a separation can be to include a naphthyl or
anthryl (electron-rich) group in the stationary phase
to cause it to interact *via* the *pi* bonds with an
electron-deficient aromatic moiety (e.*g*. dinitrophenyl)
in the solute, perhaps after suitable derivatization of
the latter (or *vice versa*). Similarly, the hydrogen
atom of a secondary amide can interact strongly with
the oxygen of a carboxyl or silanol group *via* hydrogen
bonding, and this again is a guide to the most suitable
derivative. For optimum resolution, the asymmetric
carbon atoms should be as close as possible to the
atoms or groups which interact.

Application of these principles to some selected
lipids is described in the following sections.

2.   RESOLUTION OF MONO- AND DIACYLGLYCEROLS BY HPLC

Separation of Diastereomeric Derivatives on Silica Gel

An important stereochemical problem in glycerolipid
chemistry is the resolution of diacyl-*sn*-glycerols,
which are biosynthetic precursors of triacyl-*sn*-
glycerols and glycerophospholipids. One approach has
been to react diacylglycerols with isocyanates prepared
from enantiomerically-pure amines to form
diastereomeric urethane (or carbamate) derivatives of
which the most useful have proved to be (*S*)-(+)- or
(*R*)-(-)-1-(1-naphthyl)ethyl urethanes.

The required reagents are available commercially and
reaction with alcohols occurs rapidly, without
racemization, in the presence of a basic catalyst. The
aromatic moiety facilitates detection by UV absorbance
(at 280 nm).

Diastereomers of this kind can be resolved by
adsorption chromatography on silica gel, because they
have different physical and chemical properties. Thus a

chiral derivatizing agent (R)-X will react with a
racemic substance (R,S)-Y -

$$(R)\text{-X} + (R,S)\text{-Y} \text{ ---> } (R)\text{-X-}(R)\text{-Y} + (R)\text{-X-}(S)\text{-Y}$$

The degree of separation of the two diastereomers in a
chromatographic system will depend on the chiral
structures of X and Y and the manner of their
interactions with the mobile and stationary phases. In
addition, it is noteworthy that the order of elution of
the diastereomeric derivatives is reversed if the other
enantiomer of the reagent, *i.e.* (S)-X, can be employed.
It appears that the presence of the hydrogen atom on
the nitrogen atom between the chiral centres is
essential to the separation process and is presumed to
be a primary site for hydrogen bonding to silanols on
the adsorbent surface.

    Diastereomeric urethane derivatives were first used
for HPLC resolution of chiral glycerol compounds by
Michelsen *et al.*, but the practical value was limited,
since it was only possible to effect separation of a
restricted range of alkylglyceryl ethers; partial
separation of homologues caused groups of components of
different stereochemistry to overlap when the method
was applied to mixed substrates.[7]

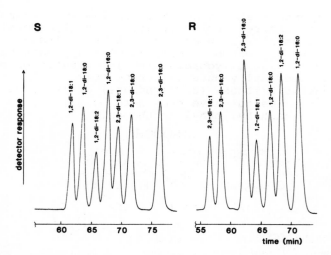

Fig. 1. Separation of a standard mixture of single acid
1,2- and 2,3-diacyl-*sn*-glycerols in the form of the
chiral 1-(1-naphthyl)ethyl urethane derivatives by
HPLC. S: (S)-(+)-1-(1-naphthyl)ethyl urethanes. R: (R)-
(-)-1-(1-naphthyl)ethyl urethanes.[8] Two columns of
silica gel (Hypersil™ 3μm, 250 x 4.6 mm i.d.) in
series were used with hexane-isopropanol (99.5:0.5,
v/v) at a flow-rate of 0.8 mL/min as the mobile phase
and UV detection at 280 nm. (Reproduced by kind
permission of the authors and of *Lipids*).

In contrast, Laakso and Christie[8] obtained an excellent resolution of diastereomeric naphthylethyl urethane derivatives of diacyl-*sn*-glycerols by HPLC on silica gel; diacyl-*sn*-glycerols derivatized with (*S*)-(+)- 1-(1-naphthyl)ethyl isocyanate eluted in the order 1,3-, followed by 1,2- and 2,3-isomers. Molecular species of single-acid diacyl-*sn*-glycerol derivatives were also resolved in the order 18:1 < 18:0 < 18:2 < 16:0, *i.e.* neither that expected for normal- nor reversed-phase partition chromatography, as illustrated in Figure 1. When the (*R*)- form of the derivatizing agent was used, the elution order changed as expected, so that 2,3- emerged before 1,2-isomers. If the two forms of the derivatives could be utilized in conjunction with HPLC-mass spectrometry, it should be possible to analyse complex mixtures in some detail. Initially, hexane with a small amount of propan-2-ol was used as mobile phase, but it was later shown that slightly better resolution and more reproducible retention times were obtainable with propan-1-ol (containing 2% water) in hexane or isooctane.[9]

This separation served as the basis for a new procedure for the stereospecific analysis of natural triacyl-*sn*-glycerols, *i.e.* for the determination of the compositions of the fatty acids esterified at each of positions *sn*-1, *sn*-2 and *sn*-3.[8,9]

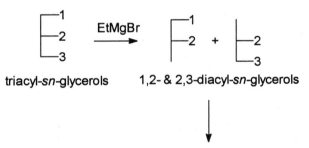

triacyl-*sn*-glycerols   1,2- & 2,3-diacyl-*sn*-glycerols

(1) preparation of (*S*)-naphthylethylurethanes

(2) separation by HPLC on silica gel

(3) methylation and GC analysis

Schematic procedure for stereospecific analysis
of triacyl-*sn*-glycerols

In the first step, triacyl-*sn*-glycerols were subjected to partial hydrolysis with ethyl magnesium bromide and the mixed products were converted immediately to (*S*)-(+)-1-(1-naphthyl)ethyl urethane derivatives. The required diacylglycerol derivatives were isolated by solid-phase extraction chromatography on octadecylsilyl columns, and the diastereomeric forms

were resolved by HPLC on silica gel. Following precise
determination of the fatty acid compositions of the
intact triacylglycerols and of the 1,2- and 2,3-diacyl-
sn-glycerol derivatives, the compositions of positions
sn-3 and sn-1 respectively could be calculated; that of
position sn-2 was obtained by difference. The procedure
has now been adapted for small samples.[10] In the past,
this type of analysis could only be achieved by time-
consuming procedures involving synthetic steps,
hydrolyses with stereospecific lipases and more complex
separation procedures. Some comparable approaches to
the problem are described below.

Diacylglycerols in the form of R-(+)-1-phenylethyl
urethane derivatives have also been resolved by HPLC on
columns of silica gel in studies designed to determine
the positional specificity of lipases with natural
triacyl-sn-glycerols.[11,12]

Separation of Enantiomers on Chiral Stationary Phases

In recent years, great advances have been made in
the resolution of chiral glycerolipids by HPLC,
especially in the laboratory of Takagi in Japan. The
approach has been to prepare 3,5-dinitrophenyl urethane
(DNPU) derivatives of mono- and diacyl-sn-glycerols and
related compounds for resolution by HPLC

on columns containing a stationary phase with chiral
moieties bonded chemically to a base of silica gel.
Once again, the presence of the hydrogen atom on
nitrogen in the urethane group, which is available for
hydrogen bonding with the stationary phase, appears to
be essential for resolution. The 3,5-dinitrophenyl
moieties of the urethanes, as well as being UV-
absorbing for detection purposes, contribute to charge-
transfer interactions with functional groups having *pi*
electrons on the stationary phase, such as that
illustrated (Sumipax[TM] OA-4100).

For example, enantiomeric monoacylglycerols were first separated after conversion to the 3,5-dinitrophenyl urethanes on an HPLC stationary phase consisting of (S)-2-(4-chlorophenyl)isovaleroyl-D-phenylglycine linked to silica (Sumipax[TM] OA-2100).[13,14] Not only were enantiomers separated, but some separation of homologous compounds was achieved such that parallel linear relationships were observed between the logarithms of the retention volumes and the carbon numbers of the compounds for the sn-1 and sn-3 series; retention times decreased with increasing chain-length. Greatly improved separations were reported when the stationary phase illustrated, containing N-(R)-1-(1-naphthyl)ethylaminocarbonyl-(S)-valine as the functional moiety bonded to silica gel (Sumipax[TM] OA-4100), was used.[15] By employing a long column, slow flow-rate, low temperatures and mobile phase of low polarity, and accepting very long retention times as a result, base-line separations of homologues differing in total chain-length by as many as ten carbons were achieved within each group of enantiomers.

This methodology has also been adapted to serve as the basis of a procedure for stereospecific analysis of triacyl-sn-glycerols, even of such difficult samples as fish oils.[16-19] In this instance, 1 and 3-monoacyl-sn-glycerols were prepared from triacylglycerols by partial hydrolysis with ethyl magnesium bromide for resolution as the dinitrophenylurethane derivatives by chiral-phase HPLC, as illustrated in Figure 2. The positional distributions in each of positions sn-1, -2 and -3 could be calculated by direct analysis of the products. The procedure has recently been used by others for a seed oil.[20]

Parallel separations of 3,5-dinitrophenyl urethane derivatives of natural chiral diacyl-sn-glycerols and structurally-related compounds by HPLC on chiral phases have also been accomplished. Clear resolution of single-acid 1,3-, 1,2- and 2,3-diacyl-sn-glycerols (and analogous ether lipids) was achieved and straight-line relationships between the logarithms of the retention volumes and carbon numbers were observed for homologous series of each form, especially with the Sumipax[TM] OA-4100 column.[21-23] Molecular species differing by two carbon atoms were separated to the base-line and distinct peaks were obtained for homologues differing in total chain-length by up to six carbon atoms. As with the monoacylglycerol derivatives described above, a long column, slow flow-rate and mobile phase of low polarity were utilized for optimum resolution. Recently, it has been established that temperatures as low as -30°C greatly improve resolution.[24-26]

A column in which a polymer of (R)-(+)-1-(1-naphthyl)ethylamine moieties was bonded chemically to silica gel (YMC-Pack A-K03[TM]) has been applied to the

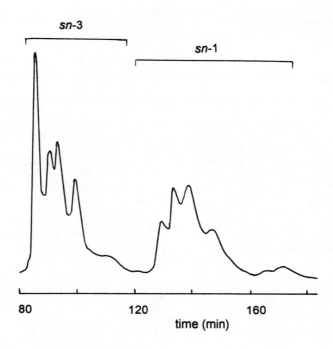

Fig. 2.   HPLC of 1-monoacylglycerols, formed by partial
hydrolysis of herring oil triacyl-*sn*-glycerols with
ethyl magnesium bromide, as their di-3,5-
dinitrophenylurethanes derivatives on a chiral column
OA-4000 (500 x 4 mm i.d.); mobile phase, hexane-1,2-
dichloroethane-ethanol (40:12:3 by volume) at a flow
rate of 0.5 mL/min at -7°C with detection at 254 nm.[19]
(Reproduced by kind permission of the authors and of
the *Journal of the American Oil Chemists' Society*, and
redrawn from the original).

resolution of diacyl-*sn*-glycerol derivatives in a
similar manner.[27] The method was applied to
diacylglycerol derivatives generated from natural
triacyl-*sn*-glycerols by partial hydrolysis with a
Grignard reagent, and for stereospecific analysis of
the latter.[28] In addition, it was utilized to isolate
1,2- and 2,3-*sn*-diacylglycerols so that the molecular
species of each could be determined by capillary GC on
a polar stationary phase.[29] In this instance, the DNPU
derivatives were resolved on the chiral column, before
the parent diacylglycerol molecules were regenerated by
reaction with trichlorosilane (with no acyl migration).
The methodology has been used to study the specificity
of enzymes involved in triacylglycerol
biosynthesis.[30,31]

As the columns containing the required chiral stationary phases and the derivatizing reagent become more readily available, this methodology will certainly be used more in structural studies of glycerolipids.

3. RESOLUTION OF CHIRAL OXYGENATED FATTY ACIDS BY HPLC

Separation of Diastereomeric Derivatives on Silica Gel

As with acylglycerols, natural enantiomeric hydroxy acids and eicosenoids have been resolved by HPLC on silica gel following conversion to appropriate diastereomeric derivatives, although chiral-phase chromatography is in general favoured now. There appears to be no consensus as to which derivative is best since commercial availability has imposed a restriction. For example, dehydroabietyl isocyanate cannot be purchased from commercial sources, but urethane derivatives of this have been used in a number of HPLC applications to eicosenoids, such as to epoxyeicosatrienoic acids following conversion to hydroxy compounds, and to various hydroxy-eicosatetraenoic acids *per se* (see reference 2 for further examples). Diastereomeric (R)-$(-)$-menthyloxycarbonyl and $(+)$-α-methoxy-α-trifluoromethylphenylacetate derivatives of hydroxy fatty acids derived from linoleic and arachidonic acids have been resolved by HPLC on silica gel in the same way.

Racemic hydroperoxides of polyunsaturated fatty acids have been resolved by preparing novel diastereomeric perketals, *i.e.* 2-propenyl ethers derived from *trans*-2-phenylcyclohexanol, which are stable during HPLC analysis.[32] Many positional isomers were separable by HPLC on silica gel, usually with the hydroperoxide of the *S*-configuration preceding the *R*-form. In addition, excellent resolution of diastereomers was obtained by reversed-phase HPLC on an octadecylsilyl stationary phase, but hydroperoxides of the *R*-configuration always preceded the *S*-form in this instance. The ketal derivatives could be hydrolysed back to the hydroperoxides under acidic conditions, without racemisation.

With compounds containing two chiral carbons, no derivatization is necessary in theory to obtain resolution of diastereomers and some practical examples of such separations have been recorded. For example, diastereomeric $(8S,15S)$- and $(8R,15S)$-dihydroxyeicosatetranoates have been resolved by HPLC on columns of silica gel, the stereochemistry of the double bonds also having some effect on retention times.[33] Other examples are cited in the earlier review.[2]

Separation of Enantiomers on Chiral Stationary Phases

In recent years, analysts have been turning increasingly to chiral stationary phases for HPLC in order to separate enantiomers of hydroxy fatty acids and hydroxy-eicosenoids especially. Such columns are easy to use and may give better products than alternative methods, because the commercial stationary phases are made from materials of particularly high enantiomeric purity. Although resolution may be enhanced if lipids are converted to specific derivatives prior to chromatography, this is not always necessary and it may be advantageous to eliminate derivatization-hydrolysis steps. Disadvantages are that the columns are costly and may not be reproducible in their properties, loss of resolution becoming apparent after continued use through leaching of ionically-bound species comprising part of the stationary phase.

Initially, the most widely used column of this kind comprised (R)-(-)-N-3,5-dinitrobenzoylphenylglycine (DNBPG) moieties bonded chemically to silica gel (Bakerbond™, Regis Hi-Chrom™). A column of this type was used to separate completely the enantiomeric forms of the hydroxy compounds derived from linoleic acid by the action of lipoxygenase (followed by reduction), for example, and to partially separate all the related compounds prepared from arachidonic acid in a single chromatographic run.[34] In this instance, the carboxyl group was methylated but the hydroxyl group was not derivatized. Others have subsequently resolved many other racemic hydroxyeicosenoids by the same procedure.[2]

Following conversion of hydroxy-eicosenoids to benzoyl or naphthoyl derivatives, much better resolution on columns of the DNBPG type was reported.[35] It seems probable that interactions between the electron-rich benzoyl/naphthoyl moieties of the solute and dinitrobenzoyl groups of the chiral phase facilitated the separation. (The separation principle is thus complementary to that of DNPU derivatives on a naphthoyl-containing stationary phase discussed above). After isolation of the enantiomers by HPLC, the original hydroxy acids could be recovered by mild alkaline hydrolysis without racemization. Once more, this methodology has been adopted by many others for related compounds.[2] A related type of chiral column, *i.e.* with N-(S)-2-(4-chlorophenyl)isovaleroyl-D-phenylglycine ionically bonded to silica but in a fused-silica capillary column, was employed to resolve dinitrophenyl urethane derivatives of racemic 2-hydroxy acids.[36] Apart from the fact that particularly good resolution was achieved, an economical use was made of the costly stationary phase.

It now seems apparent that racemic hydroxy-eicosenoids with underivatized hydroxyl groups are most

easily resolved on Chiralcel™ columns, where the
stationary phase is cellulose to which aromatic
moieties are linked *via* covalent bonds. Thus,
Chiralcel™ OB is cellulose trisbenzoate, Chiralcel™
OC is cellulose trisphenylcarbamate and Chiralcel™ OD
is cellulose tris(3,5-dimethyl)phenylcarbamate.

For example, a Chiralcel™ OC column with a mobile
phase of hexane-propan-2-ol (99:1, v/v) was used to
resolve racemic 12-hydroxyeicosatetraenoates, that of
the *R*-configuration eluting before the *S*-form.[37] In
contrast, enantiomers of 8-hydroxyeicosatetraenoate
were separated on a Chiracel™ OB column, the *S*-form
eluting before the *R*-form as illustrated in Figure 3.[38]

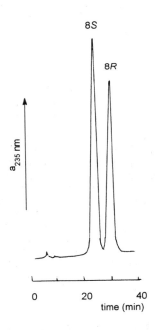

Fig. 3. HPLC resolution methyl 8*S*- and 8*R*-
hydroxyeicosatetraenoate on a column (250 x 4.6 mm) of
Chiralcel™ OB, with a mobile phase of hexane-
isopropanol (100:2, v/v) at a flow rate of 0.5 mL/min
and with UV detection at 235 nm.[38] (Reproduced by kind
permission of the authors and of *Biochimica Biophysica
Acta*, and redrawn from the original).

All the regioisomers of racemic
epoxyeicosatrienoates were resolved on columns of
Chiralcel™ OB and OD as the pentafluorobenzyl
esters.[39,40] The columns were used in the adsorption
mode with most of the isomers, but separation of the
enantiomeric 5,6-epoxyeicosatrienoates was possible in
the reversed-phase mode only, *i.e.* with a mobile phase
of water-ethanol (3:7, v/v). However, it should be
noted that great care is necessary to avoid high

pressures when Chiralcel[TM] HPLC columns are used in the
latter manner to avoid compaction of the stationary
phase.

Among recent applications, racemic
dihydroxyeicosanoates and dihydroxyeicosatrienoates
have been resolved as methyl or pentafluorobenzyl
esters on Chiralcel[TM] OC and OD columns; the diol group
was preserved intact so individual enantiomers could be
recovered for further study.[41] Similarly, a partial
resolution of related epoxy fatty acids has been
achieved on Chiralcel[TM] OC.[42] Enantiomers of 3-
hydroxymyristic acid, in the form of the N-phenylureido
diethylamide derivatives, were particularly well
separated on a Chiralcel[TM] OD column (and partially
separated by gas chromatography on a permethylated
cyclodextrin column).[43]

Appendix

When the two primary hydroxyl groups of glycerol
are esterified with different fatty acids, the lipid is
asymmetric. The enantiomers can be designated without
ambiguity by the conventional *D/L* or *R/S* systems, but
complications arise in applications to the complex
mixtures of molecular species of triacylglycerols found
in nature. A "stereospecific numbering" (*sn*) system was
therefore recommended by a IUPAC-IUB commission on the
nomenclature of glycerolipids.[44] In a Fischer
projection of a natural *L*-glycerol derivative, the
secondary hydroxyl group is illustrated to the left of
C-2; the carbon atom above this is then C-1, that below
is C-3 and the prefix "*sn*" is placed before the stem
name of the compound. Thus *sn*-glycerol-1-palmitate is
the enantiomer of *sn*-glycerol-3-palmitate

Acknowledgement

This review is published as part of a programme
funded by the Scottish Office Agriculture and Fisheries
Dept.

References

1. W.W. Christie, 'High-Performance Liquid
    Chromatography and Lipids', Pergamon Press, Oxford,
    1987.
2. W.W. Christie, 'Advances in Lipid Methodology -
    One', edited by W.W. Christie, Oily Press, Ayr,
    1992, p. 121.
3. T. Takagi, <u>Prog. Lipid Res.</u>, 1990, <u>29</u>, 277.
4. C.R. Smith, 'Topics in Lipid Chemistry. Vol. 1',
    edited by F.D. Gunstone, Logos Press, London, 1970,
    p. 277.

5. C.R. Smith, 'Topics in Lipid Chemistry. Vol. 3', edited by F.D. Gunstone, Paul Elek (Scientific Books), London, 1972, p. 89.
6. W.H. Pirkle and D.W. House, *J. Org. Chem.*, 1979, 44, 1957.
7. P. Michelsen, E. Aronsson, G. Odham and B. Akesson, *J. Chromatogr.*, 1985, 350, 417.
8. P. Laakso, and W.W. Christie, *Lipids*, 1990, 25, 349.
9. W.W. Christie, B. Nikolova-Damyanova, P. Laakso and B. Herslof, *J. Am. Oil Chem. Soc.*, 1991, 68, 695.
10. F. Santinelli, P. Damiani and W.W. Christie, *J. Am. Oil Chem. Soc.*, 1992, 69, 552.
11. E. Rogalska, S. Ransac and R. Verger, *J. Biol. Chem.*, 1990, 265, 20271.
12. E. Rogalska, C. Cudrey, F. Ferrato and R. Verger, *Chirality*, 1993, 5, 24.
13. T. Takagi and Y. Itabashi, *Yukagaku*, 1985, 34, 962.
14. Y. Itabashi and T. Takagi, *Lipids*, 1986, 21, 413.
15. T. Takagi and Y. Ando, *Lipids*, 1990, 25, 398.
16. T. Takagi and Y. Ando, *Yukagaku*, 1990, 39, 622.
17. T. Takagi and Y. Ando, *Lipids*, 1991, 26, 542.
18. T. Takagi and Y. Ando, *Yukagaku*, 1991, 40, 288.
19. Y. Ando, K. Nishimura, N. Aoyanagi and T. Takagi, *J. Am. Oil Chem. Soc.*, 1992, 69, 417.
20. S.L. MacKenzie, E.M. Giblin and G. Mazza, *J. Am. Oil Chem. Soc.*, 1993, 70, 629.
21. T. Takagi and Y. Itabashi, *Lipids*, 1987, 22, 596.
22. T. Takagi, J. Okamoto, Y. Ando and Y. Itabashi, *Lipids*, 1990, 25, 108.
23. T. Takagi and T. Suzuki, *J. Chromatogr.*, 1990, 519, 237.
24. T. Suzuki, T. Ota and T. Takagi, *J. Chromatogr. Sci.*, 1992, 30, 315.
25. T. Takagi and T. Suzuki, *J. Chromatogr.*, 1992, 625, 163.
26. T. Takagi and T. Suzuki, *Lipids*, 1993, 28, 251.
27. Y. Itabashi, A. Kuksis, L. Marai and T. Takagi, *J. Lipid Res.*, 1990, 31, 1711.
28. L.-Y. Yang and A. Kuksis, *J. Lipid Res.*, 1991, 32, 1173.
29. Y. Itabashi, A. Kuksis and J.J. Myher, *J. Lipid Res.*, 1990, 31, 2119.
30. R. Lehner and A. Kuksis, *J. Biol. Chem.*, 1993, 268, 8781.
31. R. Lehner, A. Kuksis and Y. Itabashi, *Lipids*, 1993, 28, 29.
32. N.A. Porter, P. Dussault, R.A. Breyer, J. Kaplan and J. Morelli, *Chem. Res. Toxicol.*, 1990, 3, 236.
33. A.R. Brash, A.T. Porter and R.L. Maas, *J. Biol. Chem.*, 1985, 260, 4210.
34. L. Meijer, A.R. Brash, R.W. Bryant, K. Ng, J. Maclouf and H. Sprecher, *J. Biol. Chem.*, 1986, 261, 17040.
35. D.J. Hawkins, H. Kuhn, E.H. Petty and A.R. Brash, *Anal. Biochem.*, 1988, 173, 456.
36. T. Takagi and Y. Itabashi, *J. Chromatogr. Sci.*, 1989, 27, 574.

37. S. Kitamura, T. Shimizu, I. Miki, T. Izumi, T. Kasama, A. Sato, H. Sano and Y. Seyama, <u>Eur. J. Biochem.</u>, 1988, <u>176</u>, 7225.
38. M.A. Hughes and A.R. Brash, <u>Biochim. Biophys. Acta</u>, 1991, <u>1081</u>, 347.
39. T.D. Hammonds, I.A. Blair, J.R. Falck and J.H. Capdevila, <u>Anal. Biochem.</u>, 1989, <u>182,</u> 300.
40. A. Karara, E. Dishman, J.R. Falck and J.H. Capdevila, <u>J. Biol. Chem.</u>, 1991, <u>266</u>, 7561.
41. J.H. Capdevila,, S. Wei, A. Kumar, J. Kobayashi, J.R. Snapper, D.C. Zeldin, R.K. Bhatt and J.R. Falck, <u>Anal. Biochem.</u>, 1992, <u>207</u>, 236.
42. E.H. Oliw, <u>J. Chromatogr.</u>, 1992, <u>583</u>, 231.
43. E. Kusters, C. Spondlin, C. Volken and C. Eder, <u>Chromatographia</u>, 1992, <u>33</u>, 159.
44. IUPAC-IUB Commission on Biochemical Nomenclature <u>Biochem. J.</u>, 1967, <u>105</u>, 897.

LC–GC Methods for the Determination of Adulterated Edible Oils and Fats

K. Grob

KANTONALES LABOR, PO BOX, CH 8030 ZÜRICH, SWITZERLAND

C. Mariani

STAZIONE SPERIMENTALE PER LE INDUSTRIE DEGLI OLI E DEI GRASSI, VIA GIUSEPPE COLOMBO 79, I-20139 MILANO, ITALY

1 INTRODUCTION

Although the situation has certainly improved over the last decade or two, there are still numerous oils and fats on the market, which are adulterated by admixture of other oils or fats or by consisting of products of lower quality than specified. Their detection, however, becomes increasingly demanding, because the analytical methods used in control are taken into consideration also by defrauders. Numerous producers have facilities which are better equipped than control laboratories and thoroughly check what the latter will be observing.

Frauds are promoted by the trend towards official and regimented methods: as these methods are specified, defrauders know the procedures by which their products might be tested, enabling them to adjust the fraud in such a way that it becomes undetectable. Regimentation may even render frauds legal: if the admixture is adjusted such that the parameter analyzed by the official method remains within the limit, the product must be accepted [1,2]. Efficient control and deterrence from committing frauds calls for ever new analytical approaches, threatening the clever defrauder by surprises.

Whereas 10-20 years ago, simple analysis of the fatty acids revealed numerous adulterants, today such frauds would be considered clumsy. Nowadays, fatty acid compositions may give cause for some doubt about an oil, but are seldom sufficiently significant to prove a fraud. Olive oils, for instance, often contain less than 10 % linoleic acid, but Tunisian olive oil may contain more than 20 %, which leaves ample room for sophistication. The same is true for triglyceride analyses.

If adulteration cannot be recognized by the fatty acid or triglyceride composition, then minor components must be analyzed. Since these are typically present at levels below 0.1 %, the analysis is complicated by the need to remove the triglycerides by a clean-up step. Classically this is performed by saponification, which, however, renders the procedure time-consuming: on average, saponification followed by thin layer pre-separation enables an analyst to carry out only 2-3 analyses per day, which is unacceptable for a laboratory performing broad food control such as that of the authors.

Three methods are summarized which greatly reduce or practically eliminate manual sample preparation and increase sample throughput possibly by a factor of ten. At the same time, two of them provide new evidence, enabling the detection of rather sophisticated frauds. All three methods involve pre-separation by HPLC and on-line transfer of the fraction of interest to GC (coupled LC-GC).

2 TOTAL STEROLS BY TRANSESTERIFICATION

Sterol analysis has a long tradition and is usually performed according to official methods. Triglycerides are removed by saponification; the nonsaponifiables (neutrals) are extracted, pre-separated, *e.g.* on silica gel (TLC), and analyzed by GC-FID. Saponification also causes the hydrolysis of esterified sterols (and other alcohols), *i.e.* the method determines the sum of the free and the esterified components.

An on-line LC-GC method produces the same results with less manual sample preparation and a greater accuracy [3]. Saponification has been replaced by transesterification, the latter no longer really being used for the removal of the triglycerides, but for the hydrolysis of the sterol and wax esters. Transesterification is faster than saponification (15 min at 25 °C), it avoids the thermal stress, and (most important) the difficult extraction of the neutrals from a concentrated soap solution (which is responsible for troublesome phase separations and a good proportion of the inaccuracy in the results). HPLC serves for the removal of the large amount of fatty acid methyl esters and for the pre-separation of the components of interest. The HPLC-UV chromatogram presents well separated peaks for the classes of components of interest (dimethyl sterols, methyl sterols, $\Delta 5$ and $\Delta 7$ sterols, and linear alcohols); it just remains to determine which window of HPLC retention times should be transferred to and analyzed by GC. Up to four HPLC fractions can be automatically transferred from a given HPLC run when using the Fisons Dualchrom 3000 instrument. Sample throughput has, as a consequence, increased by a factor of nearly ten, and standard deviations over the whole analysis have dropped to 1-2 %, which is better than from classical GC injection analysis.

3 DIRECT ANALYSIS OF THE MINOR COMPONENTS

The direct method analyzes minor components without prior hydrolysis, *i.e.* distinguishes between alcohols and esters. This provides additional information and simplifies the analysis. By "minor components" the following components are meant: the sterols, methyl sterols, dimethylsterols as well as their esters, the triterpenediols, the fatty alcohols, wax esters, squalene, and the tocopherols.

The method exploits the fact that many minor components are eluted from silica gel LC columns before the triglycerides, *i.e.* that they can be isolated from the oil or fat without modification (saponification or transesterification) of the glycerol esters. Free alcohols (including the sterols), however, are poorly separated from the glycerides. They must be derivatized in order to reduce their retention time in LC and to transfer them into the same region of a liquid chromatogram as the sterol esters, wax esters, and the less polar components (Figure 1). Hence free alcohols are derivatized in the oil. Then the minor components are isolated from the fat or oil by LC and transferred to GC.

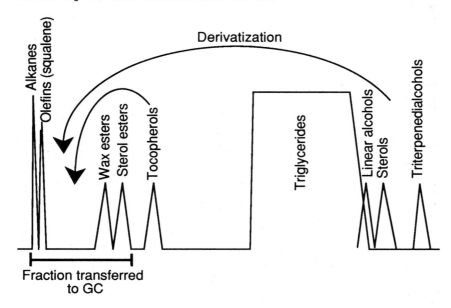

Figure 1 Hypothetical normal phase liquid chromatogram explaining the principle of the LC–GC method for direct analysis of the minor components: derivatization transfers the free alcohols into the group of the relatively apolar components. The fraction transferred to GC by the silylation method is indicated.

The first automated LC–GC method (1989) involved transformation of the free alcohols into esters of pivalic acid [4]. The bulky ester group moved the retention times of these components near to those of the sterol and wax esters of the fatty acids. A fraction just comprising the original esters and those obtained by derivatization was transferred to GC. The method was intensively used for the analysis of olive oils [5].

Later, esterification was replaced by silylation [6], which occurs at lower temperature and rules out esterifi-

cation of free alcohols by free fatty acids from highly
acidic oils. Silyl ethers are eluted from silica gel LC
columns prior to the fatty acid esters, which required an
enlargement of the "window" of the components transferred
from HPLC to GC. Additional peaks somewhat complicated
the gas chromatogram, but also increased the amount of
information obtained. A fraction was transferred which
ranged from the dead time of the HPLC column up to the
sterol esters (see Figure 1). *Mariani et al.* described a
manual method involving a classical liquid chromatographic column to achieve the same result [7].
Direct analysis of the minor components provides rich and
specific information about oils. Numerous examples have
been given [6-8], showing that usually less than 10 % of
admixed oil can be detected and identified. The minor
components in sesame oil analyzed on an apolar column
(PS-255, a methyl polysiloxane) are shown in Figure 2.

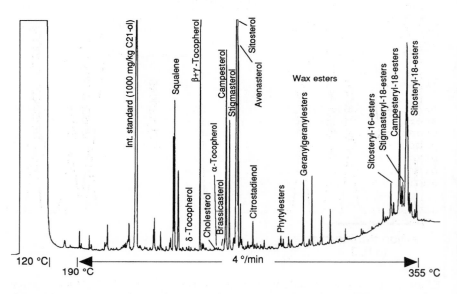

Figure 2 Direct analysis of the minor components of
sesame oil.

The tocopherols and tocotrienols provide interesting
information about many oils. A high concentration of γ-
tocopherol is typical for sesame oil. Admixture of
soybean oil, for instance, which is rather frequent, can
be recognized by the presence of δ-tocopherol. The free
sterols unambiguously demonstrate (also frequent) admixture of rapeseed oil at levels of 1-2 % by the presence
of brassicasterol. The wax esters (hardly any of which
have been positively identified) provide a fingerprint
characteristic for most oils. Also the fraction of sterol
esters is characterized by specific patterns. In the case

of sesame oil, for instance, esters of sitosterol and campesterol prevail; nearly all the stigmasterol is present as the free component.

The method also determines the wax esters, thus enabling the detection of solvent-extracted olive oil in pressed olive oil (*Mariani* and *Fedeli* [9]). It replaced the previous analysis of erythrodiol and uvaol, the concentrations of which are substantially higher in solvent-extracted oils than in pressed oils. The latter two components were usually removed, however, rendering frauds undetectable by this approach. Pressed and solvent-extracted oil from other sources can be distinguished by the concentration of the sterol esters [6].

Direct analysis of the minor components is fast and a rich source of information. In many laboratories, it has, in fact, replaced the analysis of the total sterols. There remain, however, two reasons for continuing total sterol analysis: it fits the official method in many countries (legal limits are defined for it) and enables pre-separation, *e.g.*, of sterols, methyl sterols, and dimethylsterols, as well as of Δ5 and Δ7 sterols, *i.e.* the analysis of relatively small components.

4 DETERMINATION OF THE STEROL DEGRADATION PRODUCTS

Lanzon et al. [10] proposed the analysis of dehydrated sterols for the detection of raffination or other heat treatments. Dehydration results in olefinic degradation products as shown in Figure 3 for ß-sitosterol, which primarily yields 3,5-stigmastadiene. Other degradation products are positional isomers or components with additional double bonds.

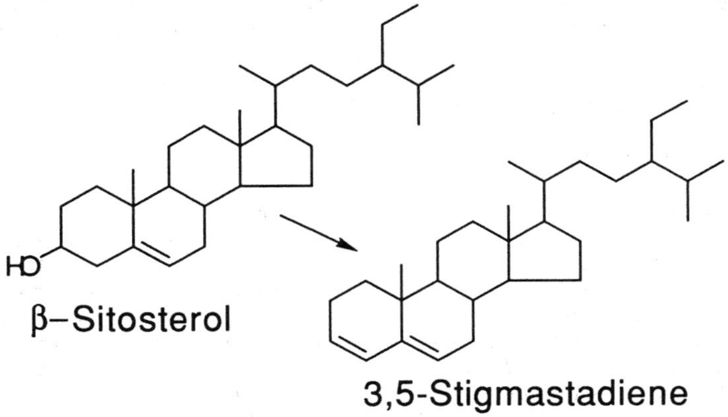

β–Sitosterol

3,5-Stigmastadiene

Figure 3 Predominant pathway for dehydration of sterols; sitosterol as example.

Analysis of olefinic degradation products enables the distinction of the standard (refined) oils from (substantially higher priced) oils labelled as, *e.g.*,

"cold pressed" or "extra virgin", or for the recognition
of admixtures of refined with non-refined oils. The conventional method for detecting heat treatment determines
conjugated dienes or trienes by UV spectroscopy. An
alternative method is, however, desirable because the UV
method is not always conclusive and techniques have been
developed which render raffination unnoticeable by UV
spectroscopy.
Conventional methods determine the olefinic degradation
products from the unsaponifiable, *i.e.* by saponification
and isolation of the olefins by TLC, LC, or HPLC. On-
line coupled LC-GC [11] renders the analysis far more
simple: the oil is diluted to a 20 % solution and transferred to the autosampler. HPLC on silica gel removes the
triglycerides, but also enables the separation of the
olefins from the alkanes and squalene. It can even be
used for the isolation of certain classes of olefins,
improving the detection of specific degradation products.
The detection limit is around 0.01 mg/kg.
Genuine cold pressed and unrefined oils and fats contain
less than 0.01 mg/kg 3,5-stigmastadiene, whereas standard
refined oils contain 3-50 mg/kg. Experiments in the laboratory and in a pilot plant confirmed that fairly drastic
conditions are required for the dehydration of the
rather stable sterols [12]. Deodoration with vapor at
temperatures up to 180 °C (1h) produced less than 0.1
mg/kg 3,5-stigmastadiene, compared with that at 200 °C of
some 0.15 mg/kg. Heat treatment of the seeds up to 150 °C
did not cause detectable dehydration of sterols. It is
concluded that the presence of olefinic degradation
products is indicative of treatments which are clearly
beyond those acceptable for cold pressed and non-refined
oils. Nevertheless 17 out of 63 olive oils sold as "extra
virgin" contained 0.2-1.6 mg/kg stigmastadiene and 24 out
of 41 other oils, labelled as non-refined, 0.3-40 mg/kg;
"cold pressed" wheat germ oils even contained up to 280
mg/kg stigmastadiene. Hence, "extra virgin" olive oils
were clearly less frequently and less severely adulterated in this respect than other oils.

Admixture of Desterolized Oils

To escape detection by control, oils for admixtures are
selected such that they do not "disturb" the fatty acid
composition. For instance, substantial amounts of high
oleic acid sunflower oil can be added to olive oils
without being detectable in this way. Admixed oils often
remain, however, easily recognizable by the sterols.
Apparently practices are used to overcome also this
problem: sterols are removed, probably by intense blea-
ching and deodoration at elevated temperature [8,13]. The
resulting frauds cannot be detected either by the fatty
acids or by the sterols.
Such sophisticated adulteration can, however, be determined by the sterol degradation products: their composition
reflects the composition of the sterols from which they
originated. As a simple example, presence of campestatri-

ene indicates the degradation of brassicasterol, which in turn shows the presence of rape seed oil. A corresponding olive oil from the market is shown in Figure 4.

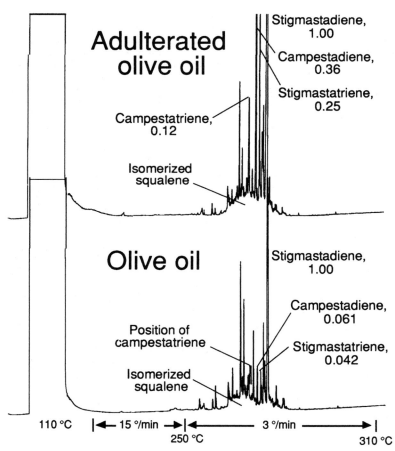

Figure 4 Olefinic degradation products of sterols and squalene in a normal refined and an adulterated olive oil. Peak areas normalized on stigmastadiene.

The lower LC–GC–FID chromatogram shows the olefinic degradation products from a normal refined olive oil. It is characterized by a large stigmastadiene peak and small peaks for stigmastatriene (from stigmasterol) and campestadiene (from campesterol, 6.1 % related to the stigmastadiene peak). The composition of the sterol dehydration products corresponds to the sterol composition of this particular olive oil and olive oils in general. An unresolved "hump" represents isomerized squalene. Some further peaks have not been identified.

The upper chromatogram of Figure 4 is from an oil, the
minor components of which fitted a refined olive oil:
there was no peak of free or esterified brassicasterol
and no increased concentrations of campesterol, stigmasterol, or their esters. The chromatogram of the olefinic
degradation products, however, no longer fits that of a
pure olive oil: it does contain the peaks of the lower
trace, confirming the presence of refined olive oil. The
sizes of the campestadiene and stigmastatriene peaks,
however, correspond to 36 and 25 %, respectively, of the
stigmastadiene peak. There is, furthermore, an additional
peak eluted at the retention time of campestatriene,
resulting from dehydrated brassicasterol. It is concluded
that this olive oil contains desterolized oils from
rape seeds and a further source contributing stigmastatriene, such as sunflower oil.

5 CONCLUSIONS

It might be impossible to achieve control capable of detecting all possible frauds, because clever defrauders
learn rapidly and develop tricks to "satisfy" the control. Progress with the three LC-GC methods mentioned
above concerns the following aspects:
1 Automated methods allow broad routine control as well
  as the analysis of a large number of control samples to
  obtain a sufficient base for comparison.
2 Direct analysis of the minor components provides a
  rather specific picture for the recognition of an oil
  (and admixtures to it). This picture can, however, be
  "corrected" by strong raffination of the admixed oil
  eliminating the characteristic components.
3 Olefinic degradation products from sterols are fairly
  sensitive indicators of raffination or other thermal
  stress, but also enable the recognition of admixed
  oils, the sterols of which have been removed.

REFERENCES

1 K. Grob, *Lipid Techn*. (in press)
2 K. Grob, H.-P. Neukom, R. Etter, and E. Romann, *Chimia*,
  1992, *46*, 420.
3 M. Biedermann, K. Grob, and C. Mariani, *Fat Sci.
  Techn*., 1993, *95*, 127.
4 K. Grob, M. Lanfranchi, and C. Mariani, *J. Chromatogr*.,
  1989, *471*, 397.
5 K. Grob, M. Lanfranchi, and C. Mariani, *J. Am. Oil
  Chem. Soc*., 1990, *67*, 626.
6 A. Artho, K. Grob, and C. Mariani, *Fat Sci. Techn*.,
  1993, *95*, 176.
7 C. Mariani, E. Fedeli, and K. Grob, *Riv. Ital. Sost.
  Grasse*, 1991, *68* 233.
8 K. Grob, A. M. Giuffré, U. Leuzzi., and B. Mincione,
  Fat Sci. Techn. (in press).

9 C. Mariani and E. Fedeli, <u>Riv. Ital. Sost. Grasse</u>,
   1986, <u>63</u>, 3.
10 A. Lanzon, A. Cert, and T. Albi, <u>Grasas y Aceites</u>,
   1989, <u>40</u>, 385.
11 K. Grob, A. Artho, and C. Mariani, <u>Fat Sci. Techn</u>.,
   1992, <u>94</u>, 394.
12 K. Grob and M. Bronz, <u>Fat Sci. Techn</u>., (in press)
13 C. Mariani, S. Venturini, P. Bondioli, E. Fedeli, and
   K. Grob, <u>Riv. Ital. Sost. Grasse</u>, 1992, <u>69</u>, 393.

Physical, Chemical, and Chromatographic Methods for the Analysis of Symmetrical Triacylglycerols – Analytical Application to an Understanding of Cocoa Butter Performance

E. W. Hammond

GROUP RESEARCH AND DEVELOPMENT, UNITED BISCUITS (UK) LTD., LANE END ROAD, SANDS, HIGH WYCOMBE, BUCKINGHAMSHIRE HP12 4JX, UK

1. INTRODUCTION

A primary ingredient of chocolate is Cocoa Butter, which is responsible for most of the functional and textural properties of a chocolate product.

As an ingredient of chocolate, cocoa butter:

suspends the other solid ingredients;
provides specific texture;
provides "coolness" and "mouthfeel";
provides significant ambient heat resistance.

These properties are very important to product quality. However, some of the excellent physical properties of cocoa butter (CB) conspire against the Confectionary Manufacturer and create high cost in process time and equipment. This is mainly because of the polymorphic nature of CB, which provides functional problems in production processes and product performance, particularly with regard to shelf life.

In most food process situations, many of the problems are effectively reduced or possibly removed by the careful determination and setting of raw material specification. To effectively do this presumes a knowledge of the raw material makeup, variability, functional properties and performance, if true success is to be achieved. The specification may also restrict the source from which the raw material is obtained, since differences in origin and crop variety can create significant changes in raw material composition and function. However, processed foods containing fats (particularly confectionary) are rarely static systems. The fats present are mixtures of solid and liquid phases. Liquid phase migration and entropy changes due to crystal morphology are readily demonstrated during the shelf life of a product. Some of these may be beneficial but are more likely to be detrimental to product performance and consumer acceptability over a period of time.

There is clearly a need to understand the performance of CB as a raw material in the process and on (or in) the product. There is also a requirement to improve its specification. These will involve an analysis of CB composition and structure. It can be shown also, that certain minor components influence the crystallisation behaviour of CB and therefore some information about the content of these may be important. This analytical information must be related to the physical performance and change of the CB under typical process and product storage conditions.

2. PHYSICAL MEASUREMENTS

Cocoa butter shows polymorphic behaviour, that is it can exist in a number of different crystalline forms. CB has a relatively simple composition (as shown in table 1). The triglycerides also demonstrate a symmetry of fatty acid composition. As a consequence of this "simple" structure and the symmetry, CB shows at least six different crystal forms which can be characterised by X-ray diffraction (figure I). Form 1 is the least stable polymorph, occurring only under conditions of rapid cooling from a melt. The stabilities of the other forms increase up to form VI, with a resulting increase in melting point. The product Technologist is concerned for CB to attain form V using tempering conditions where form V becomes stable in the product and does not change to form VI. Form V typifies good product gloss and fine texture, while form VI normally characterises "bloom".

A melted fat will tend, on allowing it to cool, to move towards its lowest state of entropy. The rate at which this happens will partly be dependent upon:

fatty acid composition;
triglyceride structure;
minor components such as diglycerides;
the temperature and heat capacity of the surroundings;
the rate of heat loss from the bulk of fat.

Where CB is concerned, the process conditions control the rate of crystallisation such that form V is seeded and induced in the product. In the product process we attempt to reach an optimum entropy state in one process pass, thereby eliminating metastable states. This hopefully leads to a quality product which changes very little during its shelf life. However, it is often possible to show that changes continue in the product. These can be accelerated by the migration of non-CB fats from other ingredients and deviating (cycling) storage temperatures, particularly where these can rise above 30°C during part of the cycle. Both these situations can alter the ratio of liquid/solid fat; in the former case probably permanently and the latter temporarily. An increase in the liquid phase due to increased

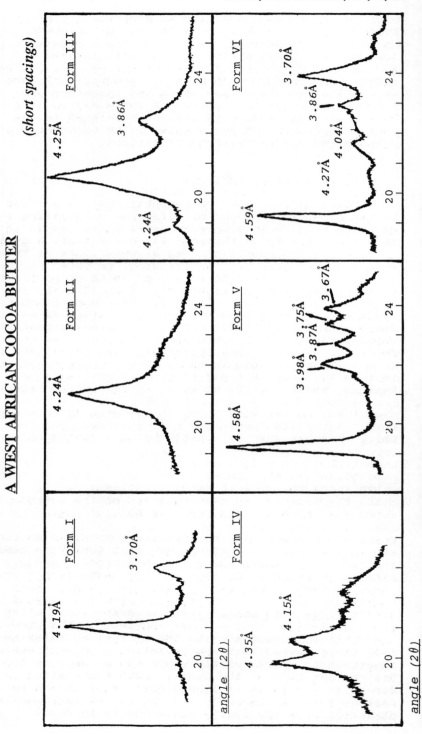

Figure 1

temperature will help migration and possibly increase
its rate.   The physical state and behaviour of CB are
therefore very important.

Table 1   Composition of a Typical Cocoa Butter

Fatty acid	%	Triglyceride	%
Myristic - C14:0	0.4	POM	1
Palmitic - C16:0	26.0	POP	17
Stearic -- C18:0	34.6	POSt	41
Oleic ---- C18:1	35.0	StOSt	27
Linoleic - C18:2	3.0	AOSt	3
Arachidic C20:0	1.0	POO/StOO	2
		PLnP/PLnSt/StLnSt	9

P = palmitic   St = stearic   O = oleic
A = arachidic Ln = linoleic

2.1. X-ray Diffraction

X-ray diffraction (XRD) while being an expensive
tool, nevertheless contributes a certain exactness to
these studies[1].   This is because it shows absolutely
which of the six crystal states[2] are present (figure 1)
at a given time.   Moreover, in making the measurement it
does not destroy the sample history.   This can be
important in shelf life studies since the same samples
can be followed right throughout the study.   It is
possible also, to control the temperature of the sample
stage in the diffractometer and to programme it from at
least -30°C to well over the melting point of CB.   Some
intricate studies involving temperature and time are
therefore possible, which provide certain information on
rate of change of crystal morphology.   However, problems
exist where the studies are to be done on whole
chocolate material or CB mixed with sugars.   The sugar
shows very strong diffraction patterns which effectively
mask the rather weaker and slightly amorphous signals
from the crystalline fat.   So these studies are in the
main restricted to whole fat samples and it becomes
difficult, or impossible, to study the nucleation
effects of crystalline sugars on CBs using XRD.

2.2. Pulsed Nuclear Magnetic Resonance

While the melting point of CB is just below human
body temperature, it is the melting curve of CB which is
important to product quality and consumer acceptance.
The melting curve has a rather special shape and it
determines hardness, "coolness" and rate of flavour
release and the luxurious mouthfeel, ending in good
clearance from the mouth with no waxiness.

CB, like other natural fats, is a mixture of
triglycerides which have different melting points, thus
in any sample at a given temperature there will be a
measurable ratio of solid to liquid fat. These ratios
when measured at specific and accurately controlled
temperatures create a percent solid fat line. The
measurements are done using pulsed Nuclear Magnetic
Resonance (pNMR) and the "N" values for a range of
temperatures can be routinely specified in raw material
specifications. Figure 2 shows a typical curve for CB
compared to those for a typical dough and cream fat. By
comparing the shape of CB curves and the extent of the
values for different CBs, we achieve a measure of their
suitability and quality for a product application. It
is very important in these studies that the conditions
used to stabilise the fat samples are constant and
accurately controlled. It is quite easy to alter the
shape of melting curves by inconsistent and non-standard
experimental procedures. Typically the methods are
based upon Waddington[3].

The pNMR technique can also be used to study what
happens when intentionally added fats (eg. butter fat)
and non-intentionally added fats (fats migrating from
another phase) are present. The additive fats may cause
small but significant changes to the melting curve. In
severe situations gross changes may occur due to
eutectic formation. By measuring the "N" values for a
series of known mixtures the pNMR studies produce
"Isosolid" diagrams (see example, figure 3). These
isosolid diagrams can be used to help define the useful
range of addition for an intentionally added fat. The
diagrams may also show any negative or positive effects
of minor components where these may be present or are
added in a similar controlled manner.

2.3. Differential Scanning Calorimetry

The rate of fat crystallisation and crystal
transformation are important when producing chocolate
confectionary. Heat energy is created and lost during
fat crystallisation. This has to be removed at a
sufficient rate to induce stability in the product and
create a state which is amenable to packaging without
product damage.

I mentioned above the need to stabilise CB in
"form V". Chocolate coated, or enrobed, products are
usually passed through a cooling tunnel prior to
packaging. This tunnel serves to assist the removal of
the heat energy produced by the crystallising CB.
However, this step is not as simple as it may appear. A
product of good or bad appearance and bloom potential
can readily be produced by correct or incorrect
profiling of the temperature along the tunnel. The
problems emanate from the way in which the CB is induced

Figure 2. Solid Fat by pNMR
Typical Fat Profiles

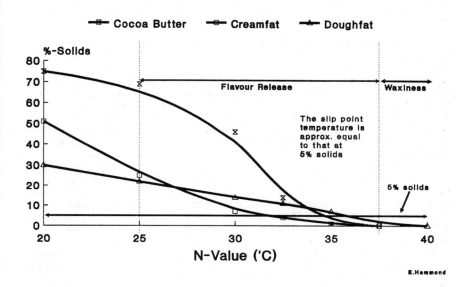

E.Hammond

Figure 3. Isosolids by pNMR

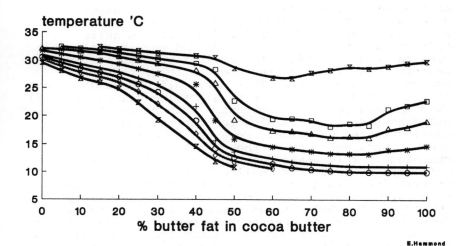

E.Hammond

to crystallise compared to its potential to crystallise
and stabilise in the required form V. Trial and error
are not very useful when rapid adjustments of a factory
cooling tunnel are required. Differential Scanning
Calorimetry (DSC) can provide a rapid and accurate tool
for these studies.

DSC can be used in a number of ways to look at the
thermal changes in CB as it crystallises. When fully
melted CB is "crash" cooled to -30°C using liquid
nitrogen it crystallises in an alpha state, normally
referred to as form I. Programming the temperature of
this "frozen" state upwards to 40°C at a slow but
constant rate, allows the CB to progress through a
series of crystal transformations. These represent the
various forms I through to VI[4]. The exothermic and
endothermic events, where these are separated and
independent, are readily measured using DSC (a Perkin
Elmer DSC7 was used in this work). Interpretation of
these thermograms is not simple and straightforward and
requires some experience. However, much very useful
information can be obtained. The peak occurring between
30°C and 35°C is an important one and is generally
believed to be associated with form V crystal.
Comparing this region in various CBs, leads to a measure
which can be used to determine variability (and possible
suitability) of CB raw materials. Observing the same
thermal event in a different way with DSC, can yield
rapid and important information about the factory
process. Rapid adjustments can be made and tested for
effectiveness.

The potential for the formation of form V in this
method can be used to aid selection of raw material CB.
Figure 4 shows a typical thermogram for two CBs with
only slight differences in their chemical composition.
Note the relatively large difference in the two
profiles. The two CBs will and do perform quite
differently in a factory process and require different
cooling tunnel profiles for optimum process performance.
However, the actual differences in their fatty acid
composition are small. This small difference does lead
to a significant change in triglyceride composition and
this is perhaps more important. This particular form of
DSC measurement is too slow for the setting up of
cooling tunnel profiles. However, DSC allows
measurements not just on the CB but also on the whole
chocolate. This means that chocolate can be sampled at
different parts of the cooling tunnel and the DSC
profile determined using a more rapid DSC procedure. In
this way optimum process conditions can be quickly
achieved. The DSC instrumentation is sufficiently
"portable" to allow transport between different factory
units. The computer control allows the setup and use of
the DSC by relatively unskilled staff. This is provided
that the refrigerated cooler is used and not liquid
nitrogen.

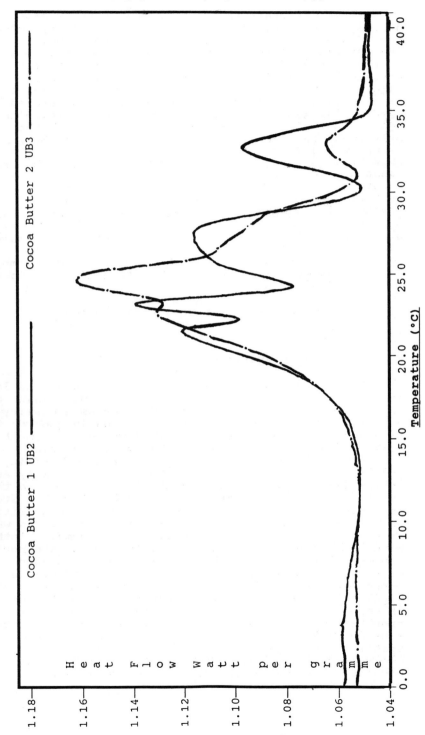

Figure 4

DIFFERENTIAL SCANNING CALORIMETRY CURVES
COCOA BUTTERS OF SLIGHTLY DIFFERENT COMPOSITION

Cocoa Butter 1 UB2 ———

Cocoa Butter 2 UB3 —·—·—

It is important to remember that the DSC experiment destroys the sample history. Therefore, in any longer term study replicate samples may have to be made up in the DSC pans. Problems exist here as you have to assume that each sample acts or reacts identically during storage and this is not definite with such small aliquots. Multiple sample analysis at each time point is therefore recommended. Sub-sampling from stored larger bulk samples can be better. However, even here you may have to take account of differences in the changes at the surface and the centre of such bulk stored samples. The pressures and thermal characteristics can often be different between centre and edge.

Differences in the crystallisation behaviour between CBs can be quite large. It is this point which suggests that often very tight compositional specifications are needed. The chromatographic analysis of CBs may produce only very small or even subtle differences between batches, whereas, the DSC thermogram can show large shifts in profile. An understanding of the relationship between chemical composition and physical behaviour can lead to a more simple and often more rapid chromatographic analytical approach.

3. CHEMICAL ANALYSIS

CB for use in chocolate manufacture should be low in free fatty acid (typically <0.5%). Iodine value may also be quoted, but is of limited value as a quality determinant. Blend ratios are also usually quoted; the ratios of CBs from different origin which make up the bulk. However, work suggests that more specific compositional data would improve the reproducibility and lead to better specification. For instance, fatty acid data will show whether one batch might be expected to be slightly softer or harder than another. The real story is again not that simple. The triglyceride composition, as shown in table 1, will be very important to the way in which a particular CB crystallises. Normally, other minor components, particularly diglycerides, are not analysed for, but they can have large effects upon crystallisation behaviour. These minor diglyceride components are present by virtue of triglyceride lipolysis during collecting, storing and processing of the cocoa beans. On delivery of the refined CB, the free fatty acid will have been removed almost entirely and the permitted level is usually quoted for in Specifications. But the same is not true for diglycerides which can reach relatively high levels (as high as 10% has been found) in some cocoa butters.

3.1. Chromatography

The chromatography of fatty acids and triglycerides has been covered in full by Hammond[5] and provides techniques which yield detailed, accurate and relatively fast information. The fatty acid composition (as fatty acid methyl esters - FAME) by gas chromatography (GC) provides a useful first point analysis for comparisons (typical data, table 1) and a very simple chromatogram (see figure 5). This may be combined with a standard GC carbon number analysis of the triglycerides (figure 5). The carbon number analysis yields a chromatogram with essentially three major peaks (C50, C52, C54) and four minor peaks (C48, C56, C58, C60). This information is of only limited use when considering the suitability of CBs since the peaks are composed of groups of triglycerides, where the sum of carbons in the three fatty acid chains equal the carbon number. This gives no information about the real distribution of saturated and unsaturated fatty acids in symmetrical and non-symmetrical triglycerides.

At the start of section 2 on physical measurements, I mentioned the importance of the symmetrical nature of CB triglycerides in providing the exceptional degree of polymorphism. These structures are also responsible in the main for the melting behaviour of CB and therefore its value as a confectionary fat. The main symmetry is about oleic acid on position 2 of the triglyceride. Positions 1 and 3, in these symmetrical types, are occupied mainly by palmitic (C16:0) and stearic (C18:0) acids and a small amount of arachidic acid (C20:0). The true distribution of fatty acids on position 2 can be obtained relatively easily, by lipase hydrolysis of the triglyceride and GC analysis of the 2-monoglyceride (method available on request). It becomes obvious now that to understand the variability of CB we require to know, at the least, the total amount of symmetrical triglyceride. This information can be obtained by argentation high performance liquid chromatography (HPLC) or plain silica HPLC analysis after epoxidation of the triglycerides[5]. A much more informative picture is obtained by separating the various symmetrical and non-symmetrical triglycerides by reversed phase HPLC (RP-HPLC; see figure 6). The distribution of carbon number and unsaturation as obtained by RP-HPLC can be cross checked with total FAME and lipolysis data to ensure an accurate picture for research purposes. The RP-HPLC procedure is routine (although requiring some skill in operation) and easily applied to a large number of production samples.

The analysis of diglyceride is relatively simple but requires attention to detail depending upon what information is required[5]. An approximate value for total diglyceride can be obtained from a standard GC carbon number analysis. This will not be accurate but

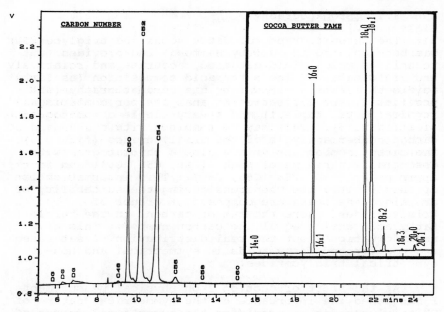

Figure 5   Carbon number analysis of Cocoa Butter by gas
chromatography. Inset is fatty acid methyl esters
by GC.

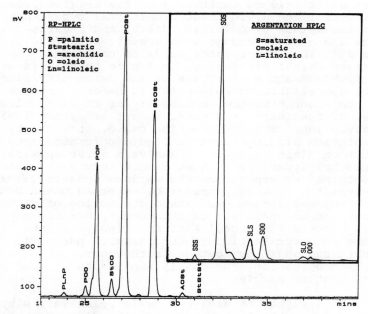

Figure 6   HPLC of Cocoa Butter triglycerides. The inset is the same
sample analysed by argentation HPLC.

can be improved substantially by converting the diglyceride to their trimethylsilyl ethers. Alternatively, total diglyceride can be obtained by HPLC without silylation. However, it may be more important to analyse the individual amounts of the two isomers 1:2 and 1:3 diglyceride at a point in time. Additionally it may be important to know the fatty acid structure and position in the diglyceride. Both these situations will require careful handling of the sample and stabilisation of the diglyceride. This is because 1:2 and 1:3 diglyceride form an equilibrium mixture, by acyl transfer, on standing in the liquid or solution state. The equilibrium is in favour of the 1:3 isomer in an approximately 80/20 ratio. Acyl transfer can be stopped by reacting the free hydroxyl group to form an ester or an ether. This reaction needs to be rapid and under conditions which do not accelerate acyl transfer. The products can be analysed by GC or RP-HPLC to obtain information on their carbon number and fatty acid composition and distribution.

4. CONCLUSION

A combination of physical and chemical analytical techniques provides immense power in the determination of the structure and composition of CB and how this affects function. This knowledge allows business interests to be optimised. Specifications of raw materials become more realistic and correct, while factory process conditions are adjusted for improved product and economic benefit.

5. REFERENCES

1. U. Riiner, Lebensm. Wiss. Technol., 1970, **3**, 101-106.

2. R.L. Willie and E.S. Lutton, JAOCS, 1966, **43**, 491-496.

3. D. Waddington in "Analysis of Oils and Fats", Ed., R.J. Hamilton and J.B. Rossell; Elsevier Applied Science Publishers, London, 1986, pp.341-400.

4. A.G. Dodson, J. Beacham, D.M. Toler, 1984, BFMIRA Research Report No.452.

5. E.W.Hammond, "Chromatography for the Analysis of Lipids", CRC Press, Boca Raton, USA, 1993.

Developments in the Analysis of Phenolic Lipids

J. H. P. Tyman

DEPARTMENT OF CHEMISTRY, BRUNEL UNIVERSITY,
UXBRIDGE, MIDDLESEX UB8 3PH, UK

1   INTRODUCTION

The phenolic lipids comprise two groups, the non-isoprenoids[1] derived from polyketides, such as the cashew phenols from the Anacardiacae (*Anacardium occidentale*), the oriental lacs from the Rhus group (*Rhus vernicifera*) and the iso-prenoids typified by the tocopherols which occur widely in many triglyceride seed oils. This review is concerned with analytical developments in the first group comprising predominantly the industrial materials technical cashew nut-shell liquid[2] (CNSL) and the lacs mainly of Chinese and Japanese origin[3]. Technical CNSL containing the C_{15} phenols, cardanol (1,a-d), cardol (2,a-d), 2-methylcardol, minor phenolic components and polymeric material is obtained by the commercial thermal decarboxylation at about 200°C of the principal natural phenolic acid component, anacardic acid (3,a-d). This is present in the shell of the cashew nut, a product of the cashew tree indigenous generally to equatorial coastal regions of Brazil, South India and East Africa. Urushiol (4,a-e), which was formerly thought to contain the 8,11,13-triene in the 11(Z) form[4,] is derived commercially by the processing[4] of the latex obtained from incisions in the trunk of the tree species *Rhus vernicifera*. Wheat and rye (*Cereale secale)* contain C_{19} to C_{25} mono and diunsaturated cardols[5,6]. The adipostatins A and B, the former of which was thought to be novel by the authors[7], although it is apparently (15:0) cardol and the latter the isopentadecyl analogue, have been isolated from an antibiotic source, *Streptomyces cyaneus*.

Although the quality control and analysis of the lipids of the oils and fats has been subject to international specifications the phenolic lipids have remained outside this scrutiny. Commodities like technical CNSL from different parts of the world have frequently been assessed by their viscosity and acidic polymerisability possibly because their chief usage has been in the polymerised form with formaldehyde as one of the main ingredients of friction dusts. As with the oils and fats, classical analytical values for technical CNSL have been described which typically[8] has an acid value (107), iodine value (296), saponification value (119) and specific gravity (1.013). Nevertheless interest both in the chemical uses of cardanol[9,10] and of problems in the polymerisation behaviour of technical CNSL towards formaldehyde[11] have made compositional analysis a desirable objective. Additionally, in such studies new components can be located[12] and structural features defined. In this account developments are described in the analysis of phenolic lipids by chromatographic, spectroscopic and combined techniques. The chromatography of natural CNSL particularly with respect to anacardic acids has been reviewed[13].

2   CHROMATOGRAPHIC METHODS

The development of methods for the analysis of technical CNSL and for oriental lacs can be grouped conveniently into thin layer (TLC) and column (CC) chromatographic techniques, gas liquid chromatographic approaches (GLC) and high performance liquid chromatographic (HPLC) procedures. These are dicussed in this order which is the chronological one. Some experimental details are given and are available fully in the references.

Adsorption Thin layer(TLC) and column chromatographic(CC) procedures for the analysis of technical CNSL and of oriental lacs

A thin layer/ ultraviolet spectrophotometric method[14] for the quantitative analysis of natural CNSL which had its origin in a purely gravimetric TLC method[15] was found difficult to apply to technical CNSL due to its dark colour, streaking and the spot width of cardanol at high R_f which was greater than the amplitude of the flying spot scanner of the Vitatron equipment. Nevertheless a purely preparative adsorption TLC gravimetric procedure operated on a number of plates with the solvent chloroform-ethyl acetate (95:5) was used for the analysis of technical CNSL[16] and with chloroform-ethyl acetate for Japanese lac[17]. Column chromatographic procedures for the analysis of technical CNSL[14] and for oriental lacs[17] in the solvent chloroform by stepwise elution with ethyl acetate proved to be useful preliminary analytical methods. Table 1 summarises the averaged % compositional results of gravimetric separations by thin layer (TLC) and column chromatography (CC) on technical CNSL (a) and Japanese lac (b). The protracted nature of these separations in which there was, at the preparative scale, only marginal resolution, although

effected at ambient temperature involved some deterioration

Table 1   Adsorption TLC and CC of Phenolic Lipids

Parameter	Cardanol	Cardol	2-Mecardol	Urushiol*	Polymer
a (TLC)	73.2	14.4	7.48	–	5.06
a (CC)	72.0	16.7	7.02	–	4.28
b (TLC)	3.3	–	–	61.9	27.1
b (CC)	4.5	–	–	58.0	17.5

(* The source of Japanese lac was the Sendai region).

of constituents. This was overcome later by the use of flash chromatography[18] which however involved the processing of many fractions. At an earlier stage by the use of hydrogenation and methylation[17], oxidative deterioration and polymerisation were arrested but quantification was less reliable because of inadequate band resolution.

Gas Liquid Chromatographic (GLC) Analysis of Technical CNSL and Oriental Lacs

Developments in the GLC analysis of technical CNSL and of lacs have consisted firstly in the analysis of the volatile total component phenols by the co-elution of saturated, monoene, diene and triene constituents, secondly in the separation of the unsaturated constituents of each component phenol following their initial TLC separation and thirdly analysis with an internal standard to determine both volatile and the non-volatile polymeric material. These stages proved inevitable because of the lack of any stationary phase at that period which could effect a total analysis.

GLC Analysis of Total Hydrogenated and Derivatised Component Phenols. The relative retention times of the total unsaturated phenols (UP), the hydrogenated phenols, the hydrogenated methyl ethers (HM) and the acetates of the unsaturated phenols (UA) of technical CNSL have been determined on the non-polar and semi-polar stationary phases 3% SE30[19], 2% and 3% PEGA[20], 3% Dexsil 300, 5% SE52, 3% OV17 and 5% APL[21] and Table 2 summarises these values which were determined with a Pye Unicam 104 chromatograph with a silanised glass column 5'x 3/16" and support with nitrogen at 45 cm^3/min. In the analysis of lacs, hydrogenated, hydrogenated/methylated, methylated and trimethylsilylated materials were employed but since only urushiol was found to be volatile the % composition was examined in the presence of an internal standard and this is discussed in that section.
Both the unsaturated acetates and phenols tended to exhibit unsymmetrical peaks due to slight resolution and were thus less amenable to straightforward triangulation procedures. PEGA although an excellent stationary phase was being used near the upper recommended thermal limit as indeed were all with the exception of Dexsil 300. Fig.1 shows typical tracings for the Dexsil and PEGA columns. The breaks in the baseline are due to attenuation changes which were shown to be linear and all indicate the separation of a C_{17} cardanol peak.

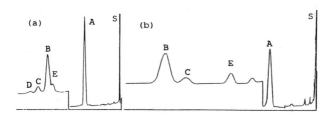

Fig.1. GLC on hydrogd./methd. technical CNSL. (a),3%
Dexsil, (b),2% PEGA. A,(15:0) cardanol Me ether, B,(15:0)
cardol Me ether, C, 2-Mecardol Me ether, D, Dimethyl
anacardate, E,(17:0)cardanol Me ether, S,solvent.

Table 2   Relative Retention Times(min.)of Component
           Phenols and Derivatives (HM,UA,UP) in
           Technical CNSL on Various Stationary Phases(SP)

SP•	Deriv.	Temp(C)	Cardanol	Cardol	2-Mecardol	A.Acid*
3%OV17	HM	220	16.62	37.96	42.24	57.48
5%SE52	HM	220	11.2	21.28	24.48	26.72
"	UA	"	15.5	40.56	45.32	–
"	UP	"	13.04	32.42	36.6	–
3%Dexsil	HM	"	14.08	28.32	31.96	34.84
"	UA	"	19.22	57.82	64.56	–
3%SE30	HM	"	12.85	22.94	26.58	–
"	UA	"	14.92	36.72	41.32	–
"	UP	"	12.68	28.38	32.58	–
3%PEGA	HM	190	19.2	61.42	53.42	–
5%APL	HM	220	36.32	74.36	82,88	–

*Anacardic Acid as the dimethyl deriv.; •SE30 and SE52
were supported on Diatomite M (100-120, BSS); OV17,
Dexsil, PEGA and APL on Diatomite C (60-80). Standard
deviations on retention determinations were   2-3%.

In the determination of the % composition of the total C_{15}
component phenols, cardanol, cardol and 2-methylcardol in
technical CNSL it was generally found most convenient to
employ the hydrogenated/methylated material and four steps
were used for the results in the analytical procedure. As
a preliminary it was essential to prepare standard mixtures
of the particular (15:0)cardanol, cardol and 2-methylcardol
derivative in order to obtain relative response factors.
Following the calculation of the uncorrected normalised %
composition from the peak areas, the results were corrected
by applying relative response factors (RRF values), the
normalised results were then expressed in terms of the
(15:0) phenols, and finally in terms of the unsaturated
phenols by the use of averaged molecular weights. Table 3
gives the % compositional results for technical CNSL
determined in this way on some of the stationary phases.
The results show reasonably good agreement and illustrate
the general principle that the % composition by GLC is
independent of the stationary phase.

Table 3 % Composition of Total Component phenols* by GLC
          of hydrogenated/methylated Technical CNSL

Parameter	Temp.(C)	Cardanol	Cardol	2-Mecardol
RRF(ArOH)	230	1.384	0.816	0.873
3% SE30	"	82.15	13.71	4.15
RRF(ArOMe)	190	1.000	0.882	0.834
2% PEGA	"	82.99	14.34	2.67
RRF(ArOMe)	220	1.000	0.882	0.834
3%Dexsil	"	83.31	13.73	2.95
RRF (UA)	"	1.000	0.768	0.644
3% SE30	"	84.01	12.17	3.80

*GLC Results for SE30 show ca. 1.07% anacardic acid and,
for 2% PEGA, 2.58% C_{17}cardanol. Standard deviations for
the component phenols were in the range, 0.33-0.58.

GLC Analysis of the % Composition of the Unsaturated
Constituents in Technical CNSL and in Lacs. The unsaturated
phenolic constituents of cardanol were found to separate on
the semi-polar stationary phase 2%PEGA but the excessively
long retention times and high temperatures involved led to
the preferred use of the methyl ethers[22]. The lack of
resolution between the constituents of different component
phenol derivatives made it essential to use a TLC/GLC
method in which the component phenols were first separated
by TLC, methylated and then examined on 2% PEGA.
Subsequently trimethylsilyl derivatives were used[23,24]. For
lacs, methyl and trimethylsilyl ethers were employed[17]. The
retention volumes of the unsaturated constituents of
technical CNSL and of Japanese lac by GLC on 2% PEGA
together with those of certain reference compounds are
depicted in Table 4.

Table 4 Retention Volumes (cm.³) by GLC of Derivatives of
          Phenols of CNSL and of Urushiol on PEGA and SE30

Parameter	SP	Temp(C)	(15:0)	(15:1)	(15:2)	(15:3)
Me Ether:						
Cardanol	2%(P)*	200	630.0	697.5	850.5	1035.0
Cardol	" "	"	-	1989.0	2407.5	2983.5
2-Mecardol	" "	"	-	1845.0	2259.0	2758.5
TMS Ether:						
Cardanol	3%(P)	190	-	430.0	505.0	609.0
Cardol	" "	"	-	750.0	882.0	1039.0
Cardanol	"(S)	200	375.0	-	-	-
Urushiol	"(S)	"	720.0	-	-	-
Phenol:						
Cardanol	2%(P)	200	3892.0	4143.0	5050.0	6173.0

*(P, PEGA: S, SE30). Ref. compd., methyl oleate, 195.0cm³.

In the lac series, methylation afforded the expected five
urushiol dimethyl ethers (DME) (Fig.2 shows a GLC tracing)
with the relative retentions on 2%PEGA: (15:0),1.00,
(15:1), 1.11, (15:2), 1.33, (15:3,(8,11,14), 1.67,

(15:3,(8,11,13),   2.43;   followed   by   the   corresponding
monomethyl ether series with the relative retentions for
the same constituents, 2.06, 2.28, 2.74, 3.47 and 2.49.
Under the same conditions the more easily formed bis TMS
ether had relative retentions repectively, 0.59, 0.65,
0.74, 0.91 and 1.32.

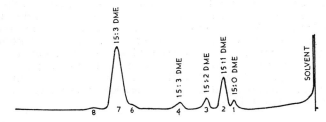

Fig.2. GLC of dimethylated Japanese Lacquer(DME)on 2% PEGA
at 200C;(2)is the 8,11,14;(7)is the 8,11,13 isomer.

The quantitative determination of the % composition of the
unsaturated constituents of the component phenols of
technical CNSL[25] followed the same methodology as for the
total phenols. The values obtained from the normalised %
composition for the four constituents were corrected by
applying relative response factors (RRF values) which were
determined from the GLC behaviour of standard mixtures of
the unsaturated constituents of cardanol and of cardol. In
the case of Japanese lac[17] it had not proved possible to
determine similar correction factors and simply the
normalised % composition was derived.
Underivatised urushiol in lacs from various regions of
Japan and China has also been investigated by capillary GLC
on a wall-coated open tubular fused-silica column treated
with a methylsilicone[26]. Urushiol conains three
stereoisomeric trienes, two dienes, a monoene and the
saturated member in the C_{15} and five compounds in the C_{17}
series. Table 5 summarises the results from technical
CNSL[25], from urushiol ethers[17] and urushiol[26].

Table 5   The %Composition of the Unsaturated Constituents
          of the Component Phenols of Technical CNSL and
          of Urushiol in Japanese Lac by GLC

Parameter	(15:0)	(15:1)	(15:2)	(15:3)	(15:3)*
RRF	1.000	0.784	0.741	0.563	-
Cardanol:	1.91	34.49	18.72	44.89	-
RRF	0.882	0.328	0.332	0.314	-
Cardol:	nil	10.74	25.71	63.55	-
2-Mecardol•	0.84	17.30	20.95	60.91	-
Urushiol**	4.93	18.45	8.85	5.78	62.00
" ••	3.83	19.44	8.79	5.79	62.13
" *	3.63	21.65	4.14	(62.34)	

*(8,11,13); **(from the diMe ether; ref.14); •(same RRF
used as for cardol); ••(from the bis TMS ether; ref.17);

•underivatised; from the Ibaraki region, (ref.26)

The similarity of the unsaturation pattern of cardol and 2-methylcardol suggests that the latter is formed by methylation of cardol[4]. The highly unsaturated nature of cardol emphasises its greater vulnerability to polymerisation and that in aged technical CNSL the proportion of cardol diminishes to a greater extent than that of cardanol. In the entry[26] for underivatised urushiol the isomeric trienes were not separated. It would seem very probable by analogy with urushiol that the triene and diene constituents in the CNSL series contain small proportions of stereoisomers. Such likely trace constituents have indeed been revealed by HPLC and suggest that capillary GLC would also separate them.

GLC Analysis of the Derivatised Phenols in Technical CNSL and in Lacs in the Presence of an Internal Standard. Both TLC and column chromatography had indicated the presence of polymeric material in technical CNSL and in oriental lacs. The quantitative determination of the non-volatile material by GLC analysis was a desirable step to obtain the total composition of both these industrially useful commodities and to aid quality control aspects. For the determination[27], hydrogenated/methylated technical CNSL was employed to avoid the possibility of further polymerisation which was believed to occur in the GLC examination of unsaturated derivatives. Accordingly the unsaturated acetates were used additionally to determine their suspected vulnerability. The retentions of a series of alkanes were examined by GLC on 3% Dexsil 300 at 230°C alongside the derivatised pure component phenols in order to select the most applicable internal standard. This information is summarised in Table 6.

Table 6   Retention Distances(RD)of (15:0) Phenolic Methyl
           Ethers, Unsaturated Acetates and C_{20} to C_{30} Alkanes

Compound	RD(mm)	RR	n-Alkane	RD(mm)	RR
Cardanol Me Ether	79.7	1.000	C_{20}	16.70	0.209
Cardol diMe Ether	161.5	2.026	C_{22}	29.85	0.374
2-Mecardol diMe Ether	184.3	2.312	C_{24}	54.33	0.682
Cardanol Acetate	116.9	1.467	C_{26}	114.60	1.438
Cardol diacetate	337.3	4.232	C_{28}	216.60	2.718
2-Mecardol "	374.1	4.694	C_{30}	411.10	5.158

From the retention information given in Table 6, the C_{28} hydrocarbon octacosane was selected as internal standard for the total compositional study of technical CNSL. For both hydrogenated/methylated and the unsaturated acetylated technical CNSL, standards were prepared containing octacosane together with the appropriate cardanol, cardol and 2-methylcardol derivative. From GLC examination of the two standards the relative molar response values (RMR values) for (15:0) cardanol, cardol and 2-methylcardol ethers and for the unsaturated acetates of each component phenol were calculated. After incorporation of the internal

standard in hydrogenated/methylated technical CNSL and in
acetylated Technical CNSL , GLC examination (on 3% Dexsil
at 230°C) afforded chromatograms from the peak areas of
which by the application of the respective RMR value and
normalisation the total % C_{15} composition was found as
depicted in Table 7.

Table 7   %Composition of Hydrogenated/methylated(HM) and
          acetylated(AU) Tech. CNSL with Internal Standard

Material	RMR	% Composition	Normalised %
(15:0)Me ether:			
Cardanol	0.785	62.86	82.50
Cardol	0.766	11.25	14.76
2-Mecardol	0.752	2.08	2.73
Polymer	–	23.81	–
Unsatd.acetate:			
Cardanol	0.596	58.59	85.64
Cardol	0.501	7.68	11.22
2-Mecardol	0.503	2.15	3.14
Polymer	–	31.59	–

It is clear that in the GLC examination of the unsaturated
acetates of technical CNSL on the stationary phase 3%
Dexsil at 230°C, polymerisation occurs to some 8% possibly
accentuated by the preheater temperature and column
temperatures being some 90°C and 40°C above the temperature
incurred in industrial processing.
The same methodology of using an internal standard for
determining a realistic % composition of oriental lacs was
applied[28] to the analysis of Japanese lac. Hydrogenated
Japanese lac with the internal standard, (15:0) cardanol
was trimethylsilylated and was examined by GLC on 3% SE30
at 200°C. Following the application of RMR values and
correction of the results it was shown that 70.80% of C_{15},
6.30% of C_{17} urushiol and polymer approximately 10.9% were
present the remaining materials consisting of cardanol,
hydroxyurushiol and bicyclic compounds.
From the outset of the present work it had been anticipated
that a total analysis of technical CNSL would prove
feasible by GLC on a semipolar stationary phase by the use
of the readily derived trimethylsilyl ethers. Although
extensive progress was made[23], incomplete resolution of all
the peaks was observed and no opportunity was available to
examine the role of capillary GC to achieve this end. With
the development of liquid chromatographic techniques the
objective of attaining total analysis by a non-thermal
method, thus precluding the occurrence of any
polymerisation, appeared an attractive alternative.

High Performance Liquid Chromatography for the total
Analysis of (a)Technical CNSL and (b) Oriental Lacs

(a) Preliminary experiments[29] on the separation of
technical CNSL on a 5μm octadecylsilane column by reversed-
phase partition conditions with isocratic elution with
methanol-water, (80:20) and UV detection enabled the four
constituents of cardanol to be separated from the three of

cardol but a more complete examination of solvent systems proved essential for separation of the constituents of 2-methylcardol and of the polymeric material. In this later work[30] gradient elution, with the solvent system, acetonitrile-water, the determination of RMR values for all the constituents and improved integration were effected. Attempts to use adsorption conditions on partisil (5μm) typically with the solvent, n-hexane-methanol (100:4) gave insufficient resolution of the constituents of cardanol. By contrast Spherisorb bonded with octadecylsilane (5μm) gave better resolution with acetonitrile-water systems than with methanol-water. Of the former, acetonitrile-water (66:34) gave an optimum separation of the unsaturated constituents of cardanol, 2-methylcardol and cardol superior to (65:35), (68:32), (70:30) or (75:25) mixtures. To effect rapid analyses taking account of all the constituents, gradient elution from acetonitrile-water to acetonitrile and finally tetrahydrofuran was found essential. Fig.3 shows HPLC tracings for two types of CNSL.

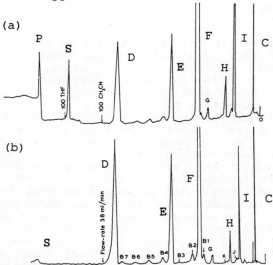

Fig.3. Reversed-phase HPLC of (a) New CNSL, (b) Distilled CNSL on Spherisorb ODS (5μm). E, cardanol monoene, D, cardanol diene, F, cardanol triene, S, satd. cardanol; I, cardol triene, H, cardol diene, G, cardol monoene; J, 2-Mecardol triene, K, 2-Mecardol diene, P, polymer, C, IS.

For quantitative work, 4-methylphenol was used as an internal standard (Is) and for the determination of the relative molar response values (RMR_p) of each phenolic constituent, a standard mixture of the unsaturated constituents and another of the saturated materials was prepared, both containing the internal standard. The Is was then incorporated in technical CNSL. With a 25x 0.4cm column, a solvent flow-rate of $1.7cm^3$/min., gradient elution and detection at 275nm, total analysis was accomplished within 60 minutes. The expression,

$$(RMR)_p / (RMR)_{Is} = (PA)_p / (g\ mole)_p \Big/ (PA)_{Is} / (g\ mole)_{Is}$$

where PA = peak area, RMR_{Is} = 1.000 and $(g\ mole)_p = w_p / Mol.Wt_p)$ was used. With this equation, knowing the other variables, w_p (the weight of the phenolic constituent) was found for each constituent and then $\Sigma(w_p)$ found which was represented as total % contribution of the weight of CNSL taken. Table 8 lists the corrected % composition for the cardanol and cardol constituents of various types of technical CNSL determined by reversed-phase (RP) partition HPLC.

Table 8   %Composition of Cardanol and Cardol in Technical
(i)new and (ii)old and Distilled CNSL by RP HPLC

Parameter	(15:0)	(15:1)	(15:2)	(15:3)	Total
(i)New CNSL:					
Cardanol,					
RMR	1.294	1.305	1.279	1.296	-
% Compn.	2.91	19.32	12.28	33.38	67.89
Cardol,					
RMR	1.001	0.997	1.003	0.993	-
%Compn.	nil	1.07	3.48	13.56	18.11
(ii)OldCNSL:					
%Cardanol	3.11	25.09	12.54	22.47	63.21
%Cardol	nil	1.06	2.57	6.59	10.22
Distilled:					
%Cardanol	3.81	30.07	16.03	32.58	82.49
%Cardol	nil	0.99	2.40	7.72	11.11

These results compare well with those determined by GLC (Table 5); e.g., normalisation of the % composition for cardol(old) from Table 8 gives (15:3),64.48%, (15:2),25.15%, and (15:1)10,37%, by comparison with 63.55%, 25.71% and 10.74% by GLC. To obtain the total corrected % composition of the three types of technical CNSL requires a knowledge of the RMR values of the remaining materials such as minor components (Fig.3, B1 to B7) and the polymer. Since the integration of trace materials proved intractable at that time and their structure was not wholly known, although they appeared to be related to $C_{15}$and C_{17}cardanol unsaturated constituents, their % contribution was determined by triangulation. The constituents of 2-methylcardol had RMR values nearly the same as those for cardol. The % polymer was then found by difference. Ideally these minor constituents should be separated by preparative HPLC and also examined by HPLC/MS to refine the total % composition. Table 9 gives a summary of the calculated % composition for three qualities of technical CNSL.

Table 9   % Composition of three Types of Technical CNSL
by HPLC Determination and by Calculation

CNSL	Cardanol (found)	Cardol (found)	2-Mecardol (calc.)	Minor (B1-B7,Calc.)	Polymer (calc.)
New	67.89	18.21	3.32	3.28	7.30
Old	63.21	10.22	1.88	3.05	21.64
Distd.	82.49	11.11	2.05	3.98	0.37

The results indicate the increase in the % polymer with ageing of the sample and its removal upon distillation. Normalisation of the % results given for the three component phenols in old CNSL, gave cardanol, 83.92%, cardol, 13.57% and 2-methylcardol, 2.50%, in reasonable agreement with values determined by GLC analysis.

(b) Following the work referred to earlier by capillary GLC on urushiol from *Rhus vernicifera* without derivatisation, concerned with the identification of stereoisomers of the trienes and dienes in the C_{15} and C_{17} series, the same group have investigated[31] the HPLC separation of urushiol diacetate from Japanese lac on silver nitrate-coated silica gel (LiChrosorb Si-60) with benzene-chloroform (4:1) into fourteen stereoisomers the total weight of which accounted for the starting solute. By analytical HPLC[32] in the reversed-phase method with acetonitrile-water-acetic acid containing silver nitrate as eluent, urushiol was separated into constituents (1)-(7) in the C_{15} urushiol series and (8)-(11) in the C_{17} laccol series also having the side-chain structures shown at the 3-position in confirmation of earlier work[26] which had demonstrated additionally the presence of (12) and (13) belonging to the thitsiol series where the side-chain occupies the 4-position. Fig. 4 shows an analytical HPLC tracing and a summary of most of the constituents from a preparative run and their structures are given in Table 10. Their ^{1}H NMR spectral properties have been described[32].

Table 10   Constituents from Prep.HPLC of Urushiol from a
                  Chinese source of *Rhus vernicifera*

Constituent	Peak,9,Fig.4	Mass, M+	%(approx.)*
C_{15}, 3-substd.			
(1) 8Z,11E,13Z-triene	C	314	64.1
(2) 8Z,11Z,14-triene	-	-	-
(3) 8Z,11E,13E-triene	E	314	-
(4) 8Z,11Z-diene	F	316	2.9
(5) 8Z,11E-diene	G	316	-
(6) 8Z-monoene	I	318	23.0
(7) Saturated	L	320	3.1
C_{17}, 3-substd.			
(8) 10Z,13E,15Z-triene	H	342	0.6
(9) 10Z,13Z,16-triene	-	-	-
(10) 10Z,13Z-diene	J	344	1.3
(11) 10Z-monoene	K	346	-
C_{15}, 4-substd.			
(12) 8Z,11E,13Z-triene	A	314	1.2
(13) 8Z,11Z,14-triene	-	-	-

    (* These %figures are from Table III, ref.26,p185.)

Similar compounds in the *Rhus* group occur in Burmese lac *(Melanorrhoea usitata)* and have been analysed[33] by mass spectroscopy (MS) and by GLC/MS, techniques which are referred to in the next section.
Reversed-phase systems[34] with octadecylsilane-silica gel columns have also been employed for the direct separation of urushiol from *Rhus toxicodendron* as well as *Rhus*

vernicifera.

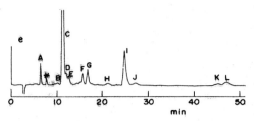

Fig.4. Reversed-phase argentation HPLC tracing of urushiol (from Hupei, China) on Develosil ODS-5 (5μm)   with UV detection (254nm) and   eluent MeCN-H$_2$O-MeCO$_2$H, (90:10:2).

3 SPECTROSCOPIC METHODS

A wide range of spectroscopic and combined spectroscopic techniques have been utilised for the quantitative and qualitative analysis respectively of phenolic lipids such as technical CNSL and oriental lacs. With cashew phenols, ^{1}H NMR spectroscopy[35,36] provides a qualitative approach to an estimation of the proportion of component phenols in mixtures and, in the separated phenols, of the proportion of individual monoene, diene and triene constituents[37]. ^{13}C NMR similarly has a potential application[37] and its value elsewhere in this book for structural assignment in poly-unsaturated lipids has been discussed by F.D. Gunstone. By mass spectrometry used in conjunction with preliminary TLC separation of the component phenols, their quantitative unsaturated composition has been derived without the need for derivatisation[11,38].   A TLC/UV procedure has been referred to[14] involving initial separation of the component phenols. FT/IR spectroscopy would appear to have a potential application and to be a neglected technique. The separation and determination of the urushiol composition of Chinese Lac has been studied[39] by GLC/MS and that of Burmese lac[33] by MS. ^{1}H NMR has been employed in an investigation of industrial Lac of Chinese origin[40,41] and UV spectroscopy for the determination of the composition of the trienes in urushiol[42].

Chromatographic/Mass Spectrometric Methods

    TLC/MS. Following adsorption TLC of technical CNSL, the constituents of the separated component phenols have been quantitatively analysed[38] by mass spectrometry through measurement of the peak heights of their molecular ions. The procedure then involved the determination of the relative response factors for the saturated and each unsaturated constituent in cardanol, cardol and 2-methylcardol to correct the observed peak heights. Since the molecular ions (M$^+$) of the triene, diene, monoene and saturated members differ by only 2 mass units, a further correction, for example of the (M+2)$^+$ peak height contribution of the triene to the (M)$^+$ peak height of the diene and correspondingly for the diene, monoene and saturated constituents, is necessary in order to obtain the three final corrected peak heights (the triene is exempt from the peak height correction procedure) from those for the respective molecular ions in the observed/recorded mass spectrum. The results for the four constituents of the C_{15} phenols of technical CNSL are given in Table 11 which also lists the semi-corrected results [for the %(M+2) contribution only] for the constituents of urushiol. In the practical operation of the method the peak heights measured were the average of six consecutive scans and the whole procedure was rapidly accomplished without the requirement necessarily for a mass spectrometer with high resolution.

Table 11   % Composition of the Constituents of the C_{15}
Phenols of Technical CNSL and of Urushiol by MS

Parameter	(15:0)	(15:1)	(15:2)	(15:3)	(15:3)*
Cardanol:					
RRF	1.9573	2.2621	2.5279	6.5557	–
% Compn.	3.94	31.97	16.12	47.97	–
Cardol:					
RRF	0.5074	2.7964	2.2612	2.5656	–
% Compn.	0.27	15.07	27.84	56.81	–
2-MeCardol:					
% Compn.	1.34	24.80	23.63	50.23	–
Urushiol:					
% Compn.	8.22	19.54	8.96	(	63.55	)

*8, 11, 13-isomer.

The results for the constituents of technical CNSL by MS compare with those obtained by GLC although with the same materials in the component phenols of natural CNSL the agreement between the two techniques was very close. For urushiol even the semi-corrected results show a reasonable level of comparability.

GLC/MS and LC/MS.   Ideally it is desirable to both obtain the % composition quantitatively and simultaneously to obtain verification of the structure of the constituent under experimentation by either GLC/MS or LC/MS. With the phenols in technical CNSL[23,] despite many endeavours, this has not yet been achieved quantitatively and the results have tended to have more qualitative compositional value. The potential value of this approach is in the elimination of the preliminary TLC separation. As in previous work, the relative response factors for the four constituents of trimethylsilylated cardanol and cardol were first obtained from the corrected peak heights of six consecutive MS runs and GLC/MS then applied to trimethylsilylated technical CNSL. The results are summarised in Table 12.

It seems probable that thermal deterioration of the vulnerable triene constituent occurred both with cardanol and cardol since the % of this is appreciably lower than with GLC analysis. The reactive cardol constituents appeared to be susceptible to deterioration in temperature-programmed GLC/MS[43] with a capillary column. LC/MS is likely to prove more useful due to the greater resolution achievable and the favourable ambient temperature conditions. There is also considerable scope for the argentation HPLC of technical CNSL in the way this has been used for Japanese lac and for the natural anacardic acids of *Pistacia vera*[44]. Its merits amongst other techniques for neutral lipids have been described in Chapter 2 of this book by L.G. Blomberg and M. Demibuker.

Table 12   % Composition of Cardanol and cardol from
            GLC/MS• on trimethylsilylated Technical CNSL

Parameter	(15:0)	(15:1)	(15:2)	(15:3)
Cardanol:				
Mol.Ion	376	374	372	370
% Compn.*	14.979	55.70	12.611	13.82
RRF	0.724	2.441	3.947	7.047
% Compn.**	3.69	46.25	16.93	33.13
Cardol:				
Mol. Ion	464	462	460	458
% Compn.*	5.436	19.111	30.606	40.449

(* Compn. corrected for M+2 peak; ** Total Correction; •
with 2% OV17 in a packed column)

ACKNOWLEDGEMENTS

Thanks are due to 3M Research Ltd. for financial assistance which made part of this work possible and for certain sources of technical CNSL, to Dr. M. Sato who supplied Japanese lac and to the Xi'an Research Institute of Lacquer more recently for Chinese material. Postgraduate research students S.K. Lam and M.A. Kashani, and final year undergraduates, K.H. Tam and A.J. Matthews who contributed to this work.

REFERENCES

1. J.H.P. Tyman, 'Studies in Natural Products
    Chemistry', Ed. Atta-ur-Rahman, Elsevier,
    Amsterdam, 1991, Vol.9, p.313.
2. E.L. Short, J.H.P. Tyman and V. Tychopoulos, J. Chem.
    Tech. and Biotechnol., 1992, 53, 389.
3. J. Kumanotani, Internat. Symp. Oriental Lacs, Fuzhou,
    China, May, 1993, Abs. 21; Proc. (in the press).
4. J.H.P. Tyman, Chem. Soc. Rev., 1979, 8, 499.
5. A. Kozubek, Z. Naturforsch., 1992, 47c, 608.
6. L.M. Seitz, J. Agri. Food Chem., 1992, 1541.
7. N. Tsuge, M. Mizokami, S. Imai, A. Shimazu and H.
    Seto, J. Antibiotics, 1992, 45, 886.
8. J.G. Ohler, 'Cashew', Royal Tropical Institute,
    Dept. Agric. Res., 1979, Amsterdam, p.76.
9. J.H.P. Tyman, 'Studies in Natural Products
    Chemistry', Ed. Atta-ur-Rahman, Elsevier,
    Amsterdam, Vol. 17, (in the press).
10. V. Madhusudhan, T. Ramalingam and B.G.K. Murthy, Eur.
    Coatings J., 1989, 6, 608.
11. J.H.P. Tyman, D. Wilczynski and M.A. Kashani, J.

Amer. Oil Chem. Soc., 1978, 55, 663.

12. J.H.P. Tyman, J. Chem. Soc., 1973, 1639.
13. J.H.P. Tyman, 'Handbook of Chromatography: Lipids',
    Eds. H.K. Mangold, N. Weber and K.D. Mukherjee,
    CRC Press Inc., Boca Raton, USA, 1992, Ch. 15.
14. J.H.P. Tyman, J. Chromatogr., 1978, 166, 159; J.H.P.
    Tyman, 'Industrial Applications of Quantitative TLC
    Analysis', Ed. L.R. Treiber, M. Dekker, New York,
    1987, Ch.5.
15. L.J. Morris and J.H.P. Tyman, J. Chromatogr., 1967,
    27, 287.
16. S.K. Lam, M. Phil. Thesis, Brunel University, 1976.
17. J.H.P. Tyman and A.J. Matthews, J. Chromatogr., 1982,
    235, 149; 26th IUPAC Congress, Tokyo, 1977.
18. J.H.P. Tyman and V. Tychopoulos, J. Planar
    Chromatogr., 1988, 1, 227.
19. J.H.P. Tyman, J. Chromatogr., 1975, 111, 285.
20. J.H.P. Tyman, Analyt. Chem., 1976, 48, 30.
21. J.H.P. Tyman, J. Oil Chemists' Soc., 1978, 55, 663.
22 J.H.P. Tyman, J. Chromatogr., 1975, 111, 277.
23. K.H. Tam and J.H.P. Tyman, (unpublished work).
24 A.P. France and J.H.P. Tyman, (unpublished work).
25. J.H.P. Tyman, J. Chromatogr., 1977, 138, 97.
26. Y. Du, R. Oshima, H. Iwatsuki and J. Kumanotani, J.
    Chromatogr., 1984, 294, 179.
27. J.H.P. Tyman, J. Chromatogr., 1978, 156, 255.
28. A.J. Matthews, J.H.P. Tyman and R.A. Stone, Internat.
    Symp. Oriental Lacs, Abstr. 39, Fuzhou,
    China, May, 1993.
29. K.H. Tam and J.H.P. Tyman, (unpublished work), 1978.
30. J.H.P. Tyman, V. Tychopoulos and B.A. Colenutt, J.
    Chromatogr., 1981, 213, 287.
31. Y. Yamauchi, R. Oshima and J. Kumanotani, J.
    Chromatogr., 1980, 198, 49.
32. Y. Du, R. Oshima and J. Kumanotani, J. Chromatogr.,
    1984, 284, 463.
33. A. Jefferson and S. Wangchareontrajkul,
    J. Chromatogr., 1986, 367, 145.
34. M.A. Elsohly, P.D. Adawadkar, C.Y. Ma and C.E.
    Turner, J. Nat. Prod., 1982, 45, 532.
35. P.Gedam, P. Sampathkumaran and M. Sivasamban, Indian
    J. Chem., 1972, 10, 388.
36. J.H.P. Tyman and N. Jacobs, J. Chromatogr., 1971, 54,
    83.
37. J.H.P. Tyman, R.J. Edwards and A.A. Najam,
    (Unpublished work).
38. J.H.P. Tyman, J. Chromatogr., 1977, 136, 289.
39. C. Wu, Z. Ceng and S. Da, Tuliano Gongye, 1981, 27,
    344; Chem. Abs., 96, 87058.
40. Q. Lin, L.Li and B. Wang, Kexue Tonbao, 1982, 27,
    344; Chem. Abs., 97, 40376.
41. H. Peng and X. Liu, Kexue Tonbao, 1983, 28, 274;
    Chem. Abs., 99, 89647.
42. W. Zhao, Y. Feng and C. Yu, Linchan Huaxue Yu Gongye,
    1982, 2, 30; Chem. Abs., 97, 211934.
43. J.H.P. Tyman and B. Wheals, (unpublished work).
44. M. Yalpani and J.H.P. Tyman, Phytochemistry, 1983,
    22,137.

^{13}C NMR of Lipids

F. D. Gunstone

LIPID CHEMISTRY UNIT, SCHOOL OF CHEMISTRY, UNIVERSITY
OF ST ANDREWS, FIFE KY16 9ST, UK

1   INTRODUCTION

During the '70s interest in the ^{13}C nmr spectra of lipids
was concentrated mainly on assigning chemical shifts for
natural and synthetic fatty acids and esters available as
pure compounds.[1-4] More recently interest has shifted to
the study of mixtures - both of natural and industrial
origin - with attempts to use the chemical shift
assignments along with intensity data to produce
quantitative or semi-quantitative information. The older
studies and the recent developments have been reviewed.[5]

Many of the results reported in this paper have been
produced in the writer's laboratory with financial support
from the Karlshamns Research Foundation. Some of the
results have not yet been published. Spectra were obtained
with a Bruker AM 300 spectrometer (pulse angle 45^0, pulse
repetition time 1.82 sec, 1.22 Hz per data point, ~1000
scans) using solutions in CDCl3.

Walnut Seed Oil

The spectrum of walnut seed oil (figure 1) is typical
of those obtained for oils and fats composed almost
entirely of triacylglycerols of the common saturated and
unsaturated acids. Along with the spectrum the
spectrometer provides a print-out of chemical shifts and
intensities for 42 signals. Not all of these are easily
seen in Figure 1. The interpretation of this spectrum
shows what information can be obtained in a typical case.
Less common features which may be superimposed on this
general pattern are discussed in later sections.
Signals for the ω1-3, C1-3, allylic, and olefinic carbon
atoms are easily distinguished and will be discussed in
turn.* Ten signals between 29.1 and 29.8 ppm relate to CH_2
groups not easily distinguished from each other (the
methylene envelope) and these are not considered further.

---

*ω1 ω2 ω3   C3 C2 C1
CH_3 CH_2 CH_2 . . . . CH_2 CH_2 COO-

<u>Figure</u>   High Resolution ^{13}C NMR spectrum of
                      Walnut Oil

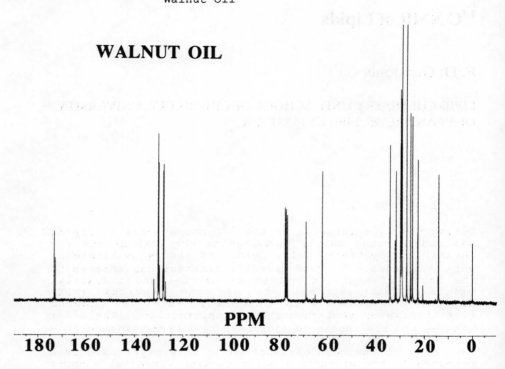

WALNUT OIL

PPM

| 180 | 160 | 140 | 120 | 100 | 80 | 60 | 40 | 20 | 0 |

Signals at approximately 173 (C1), 130 (olefinic), 77
(solvent), 69 and 62 (glycerol), 34 (C2), 32 (ω3), 29-30
(methylene envelope), 27 and 25.6 (allylic), 25 (C3), 23
(ω2), and 14 ppm (ω1).

The glycerol carbon atoms produce signals at 68.92 (β) and
62.11 ppm (α) respectively. Assignment of chemical shifts
is based on the study of individual molecular species or of
oils rich in a single fatty acid.

Signals for the ω1-ω3 carbon atoms appear as clusters
since slightly different signals are observed for *n*-3 acids
(α-linolenic), *n*-6 acids (linoleic), and for *n*-9 (oleic)
and saturated acids. The ω3 atom for *n*-3 acids is olefinic
and is not included here. Since the three ω2 signals are
resolved better than the ω1 signals they probably give
better quantitative results and, on the basis of their
intensities, indicate the percentage of the three types of
acids shown in Table 1. [Good quantitative data can only
be obtained when appropriate relaxation times are built
into the spectra-gathering parameters.] The ω2 carbon atom
for *n*-3 acids is also allylic to a *cis* double bond and thus

<u>Table 1</u>   Chemical shifts (ppm) for the ω1-3 carbon
            atoms in walnut seed oil

	ω1	ω2	ω3
n-9, saturated	14.12	22.71 (31%)	31.93
n-6	14.08	22.59 (62%)	31.55
n-3	14.28	20.57 (7%)	–

has a much lower chemical shift than the non-allylic ω2
signals.

Carbon atoms near the glycerol ester end of the chain
(C1-3) produce double signals corresponding to acyl chains
in the α and β positions (Table 2). Further differences in
chemical shifts for the C1-C3 signals of saturated and
unsaturated (Δ9) acyl chains are very small and can be seen
only when the spectra are collected over 6-24 hr.   Such
differences are more easily observed when the double bond
is closer to the acyl group as in Δ4, Δ5, or Δ6 acids.

The walnut oil spectrum shows three allylic signals.
In some spectra the signal at about 27.2 is replaced by two
signals about 0.05 ppm apart.   These are for methylene
groups α to one double bond such as O8, O11, L8, L14, Ln8
(O, L, and Ln refer to oleic, linoleic, and α-linolenic
acyl chains).   Carbon atom Ln17 is both allylic and ω2 and
has been considered along with the ω2 carbon atoms.
Signals at 25.65 (L11 and Ln11) and 25.55 ppm (Ln14) are
related to carbon atoms lying between two *cis* double bonds.
The chemical shifts are different for groups allylic to
trans double bonds.

Each olefinic carbon atom has its own chemical signal
and from an oil containing oleic, linoleic, and linolenic
acid 12 signals would be expected (ignoring small chemical
shifts for α and β chains).   We generally observed only 10
such signals because of some overlapping under the
conditions we employed to collect the spectra (Table 3).
The intensities of these olefinic signals furnish a semi-
quantitative ratio for the three olefinic acids.

<u>Table 2</u>   Chemical shifts (ppm) for C1-3 carbon
            atoms in α- and β-chains

	C1	C2	C3
α	173.18	34.03	24.89
β	172.78	34.20	24.86
diff	0.40	0.17	0.03

chemical shifts may vary as much as ±0.1 but the
differences change very little.

Table 3   Chemical shifts (ppm) and intensities of
           olefinic carbon atoms in oleate, linoleate, and
           α-linolenate.

Carbon atom	oleate		linoleate		α-linolenate	
9	129.70	2.58	[129.98	9.76]	[130.19	11.94]
10	[129.98	9.76]	128.09	9.23	127.77	2.30
12	–		127.91	9.72	128.24	2.25
13	–		[130.19	11.94]	128.29	2.12
15	–		–		127.13	1.38
16	–		–		131.93	1.52

overlapping signals are in square brackets

Oils Containing Short and Medium-chain Acids (Butter Fats and Lauric Oils)

     When a fatty acid has a sufficiently long chain the
$C1-3$ signals and the $\omega 1-3$ signals are each independent of
chain length.   Differences in chemical shift are not
expected, for example, with C_{14}, C_{16}, and C_{18} acids.
However, when the chain is short enough then the $C1-3$
signals are influenced by the end methyl group and the $\omega 1-$
3 signals are affected by the ester group. The differences
are large enough to be useful for the C_4, C_6, and C_8 acids
and it is possible to recognise fats containing butterfat
($C_4>C_6>C_8$)   or   the   lauric   oils   ($C_8>C_6$). Butanoic esters
have distinct signals for all four carbon atoms, hexanoic
esters for the C3 and $\omega 1-3$ atoms, and octanoic esters for
the $\omega 3$ carbon atom.   The figures in Table 4 are taken from

Table 4   Chemical shifts (ppm) for $C1-3$ and $\omega 1-3$ carbon
           atoms in C_4, C_6, C_8, and C_{18} triacylglycerols
           (double   signals   refer   to α and β c h a i n s
           respectively)

	$(4:0)_3$	$(6:0)_3$	$(8:0)_3$	$(18:0)_3$
C1	173.14	173.31	173.29	173.27
	172.74	172.90	172.89	172.88
C2	35.94	34.03	34.08	34.07
	36.09	34.19	34.24	34.24
C3	–	24.56	24.88	24.91
		24.59	24.92	24.94
ω3	–	31.26	31.68	31.98
		31.22	31.70	
ω2	18.37	22.31	22.63	22.73
	18.40			
ω1	13.63	13.90	14.07	14.13
	13.57			

work reported by Lie Ken Jie *et al.*[6] for single acid glycerol esters. The double signals relate to the α and β chains respectively. Figures for glycerol tristearate are included for comparison purposes.

These values have been used to interpret the spectrum of butterfat and to recognise the presence of such fat in spreads.[7]

Oils Containing Branched-chain Acids (Wool Grease)

Branched-chain acids have additional signals for the branching group and changed chemical shifts for carbon atoms around the point of branching. This may be sufficient information to identify the branched chain and to locate it in the main acyl chain. [13]C Nmr spectroscopy has been used in a study of wool grease fatty acids. This mixture contains *iso*-acids and *anteiso*-acids as well as straight-chain acids.[8]

In addition to the signals listed in Table 5 small signals at 70.49 and 25.06 ppm were associated with hydroxy acids. The remainder were satisfactorily interpreted in terms of unbranched acids (~30%), *iso*-acids (~32%), and *anteiso*-acids (~32%).

Oils Containing Cyclopropene Acids (Kapok Seed Oil)

Cyclopropene acids occur in seed oils of the Sterculiaceae, in kapok seed oil (~12%), and at a much lower level (~1%) in cottonseed oil. These acids have attracted attention because they inhibit the bio-desaturation of stearic to oleic acid. The two most common cyclopropene acids are sterculic ($\underline{n}$=7) and malvalic ($\underline{n}$=6) and these are easily recognised in the spectrum of kapok seed oil by the characteristic chemical shifts of several of the mid-chain carbon atoms.[5] (Table 6)

$$CH_3(CH_2)_7 C = C (CH_2)_{\underline{n}} COOH$$

with $\overset{\displaystyle CH_2}{\diagup \diagdown}$ bridging the two central carbons

Table 5   Chemical shifts (ppm) for branched methyl groups and ω1-6 carbon atoms in unbranched and *iso*- and *anteiso*-acids

	unbranched	iso-	anteiso-
ω1	14.14	[22.68]	11.42
ω2	[22.68]	27.98	–
ω3	31.95	39.08	34.41
ω4	–	27.45	36.66
ω5	–	–	27.14
ω6	–	–	30.06
methyl	–	[22.68]	19.23

Table 6   Chemical shifts (ppm) for selected carbon atoms
           in sterculic and malvalic acids

	sterculic	malvalic		
C7	27.37	27.43	C6	a
C8	26.01	25.96	C7	b
C9	109.20	109.11	C8	c
(CH_2)	7.42	7.42	(CH_2)	
C10	109.47	109.56	C9	c
C11	26.07	26.07	C10	b
C12	27.37	27.43	C11	a

ratio of sterculic to malvalic acid from intensity values:

    (a) 28:72 (b) 38:62 (c) 26:74

Oils Containing Epoxy Acids (Vernonia Oil and Epoxidised Oils)

    Vernonia galamensis seed oil is characterised by the
presence of a high proportion (~75%) of vernolic acid (cis-
12,13-epoxyoleic). In addition to the range of signals
similar to those observed in walnut seed oil, the vernonia
oil has large signals with unusual chemical shifts which
must be associated with vernolic acid. The influence of an
epoxide group on the chemical shift of nearby carbon atoms
(on both sides of the epoxide function) is quite large and
the following values have been quoted for a cis epoxy
group: $\alpha - 1.71$, $\beta - 2.93$, $\gamma - 0.38$.[9] These figures are
superimposed on the chemical shifts expected in the absence
of the epoxide group and the chemical shifts observed for
vernolate are compared with those for oleate and linoleate
(Table 7).

Table 7   Chemical shifts (ppm) for selected carbon atoms
           of oleate, linoleate, and vernolate in vernonia
           oil.

carbon atoms	oleate	linoleate	vernolate
8	27.20	27.20	27.41[a]
9	129.66	129.90	124.05[c]
10	129.90	128.06	132.40[c]
11	27.20	25.64	26.33[b]
12	–	127.91	57.05
13	–	130.09	56.40
14	–	27.20	27.79[a]
15	–	–	26.28[b]
16	31.94	31.54	31.77
17	22.71	22.61	22.61
18	14.12	14.08	14.01

    – these signals are part of the methylene envelope
   a, b, c these assignments could be reversed

Table 8   Chemical shifts (ppm) of epoxide carbon atoms

9,10-epoxystearate	57.18	57.13		
⌈(A)*	57.19	57.13	56.71	56.44
9,10;12,13-diepoxystearate ⌊(B)*	56.99	56.93	54.33	54.17
9,10;12,13;15,16-triepoxystearate	not fully identified			

*   these refer to two different stereoisomers.

Epoxidised oils are produced on a commercial scale mainly for use as plasticisers and stabilisers in PVC. The oils are completely epoxidised and consist therefore of acyl chains with no, one, two, or three epoxide units from saturated, oleic, linoleic, and linolenic chains respectively. Acyl groups with more than one epoxide group exist in more than one stereochemical racemic form. Epoxidised palm super-olein, soybean oil, and linseed oil have been examined. These oils are particularly rich in mono-, di-, and tri-epoxy acyl chains respectively. The strong influence of the epoxide group and the presence of stereoisomers with their own ^{13}C nmr spectra means that the number of signals is quite large (45-70 for these three epoxidised oils).

Altogether in these three oils 21 epoxide signals were observed between 58.3 and 53.2 ppm. Some of these have been assigned (Table 8). Assignments have also been made for the C1 and 2 and the w1-3 carbon atoms (Tables 9 and 10). The C1 and 2 signals indicate that saturated and epoxy acyl chains share the a position and that the b chains are entirely epoxides. Since the epoxides are produced from olefinic acids this is not a surprising result. The ω1-3 signals and their intensities can be used to give a semiquantitative analysis of the epoxidised oils. Typical results are given in the original paper.[10]

Oils Containing Δ4, Δ5, or Δ6 Unsaturation (Fish Oils, Aquilegia Oil, and Oils Containing Petroselinic Acid or γ-Linolenic Acid)

With acids having a double bond closer to the ester group than in the common Δ9 acids (oleate, linoleate, α-linolenate) it is possible to see distinctive signals among

Table 9   Chemical shifts (ppm) for C1 and C2 in non-epoxy and 9,10-epoxy esters

	C1	C2
sat (α)	173.23	34.04
epoxy (α)	173.16	33.98
epoxy (β)	172.75	34.14

(all 9,10-epoxides whether from oleate, linoleate, or linolenate give the same signals)

<u>Table 10</u>   Chemical shifts (ppm) of ω1-3 atoms in 9,10-, 12,13-, and 15,16- epoxides

	ω1	ω2	ω3
9,10-epoxide (and saturated)	14.12	22.68	31.86(31.93)
12,13-epoxide	13.99	22.58	31.68
15,16-epoxide	⌈10.61 ⌊10.49	21.24 21.16	<u>a</u>

<u>a</u>   these are epoxide carbon atoms

the C1-3 clusters. In some cases the intensities can be used to determine the content of a particular acid and to examine its distribution between the α and β chains. Acids with such double bond systems are listed in Table 11. Discussion here is limited to columbinic acid in aquilegia oil and to EPA and DHA in fish oils.

Columbinic acid gives characteristic signals for 12 of its carbon atoms: C1-3, three allylic carbon atoms, and six olefinic carbon atoms (Table 12). From the intensities of the C1-3 signals it is possible to calculate the total content of Δ5 acids (mainly, if not entirely, columbinic) and the distribution of these between the α and β chains (Table 13).

From their study of lipids from the white muscle of Atlantic salmon, Aursand <u>et al</u>[13] have shown that the distribution of EPA (and arachidonic acid, AA) and of DHA between α and β chains is best calculated using the

<u>Table 11</u>   Acids having a double bond in the Δ4, Δ5, or Δ6 positions

position of first double bond	acid structure and name	source(oil)	ref.
4	22:6 (<u>n</u>-3), DHA	fish	11-13
5	20:5 (<u>n</u>-3), EPA	fish	11-13
5	18:3 (5<u>t</u>9<u>c</u>12<u>c</u>) columbinic	aquilegia	14
6	18:1 (6<u>c</u>) petroselinic	unbelliferae	14
6	18:3 (<u>n</u>-6) γ-linolenic	⌈evening primrose ⌊borage	15

<u>Table 12</u>   Chemical shifts (ppm) for selected carbon atoms of columbinic acid in aquilegia oil

C1	173.02(α), 172.62(β)	C7	32.62
C2	33.50(β), 33.32(α)	C9 (or 5)	129.22
C3	24.70(β), 24.64 (α)	C10	128.42
C4	31.87, 31.83	C11	25.72
C5 (or 9)	129.27	C12	127.81
C6	131.00	C13	130.22

Table 13   Content of columbinic acid in aquilegia oil and
          its distribution between the α and β positions

based on:	C1	C2	C3
columbinic acid			
total	60.0	60.2	66.0
in α position	56	59	58
in β position	44	41	42

intensities of the C2 and C1 signals respectively
(Table 14). These figures show the DHA particularly is
preferably in the β position. Random distribution would
give figures of 67 and 33%.

Partially hydrogenated oils

During partial hydrogenation of liquid oils some
unsaturated centres are hydrogenated while others undergo
stereomutation and/or double bond migration. The result is
a very complex mixture of isomeric acyl esters which
presents a considerable challenge to the lipid analyst.
Total trans acids can be measured by gas chromatography or
by infrared spectroscopy but there is no simple way of
determining double bond position in a mixture of monoenes.
This requires some form of oxidative cleavage - usually
ozonolysis - followed by gas chromatographic examination of
the mono- and difunctional products. This is a laborious
procedure requiring a lot of skill and care to produce
results of even modest accuracy. ^{13}C nmr spectroscopy
offers a possible solution to this problem.[16] Useful
information is obtained through consideration of the
signals for ω1-3, allylic, and olefinic carbon atoms.

Each of the ω1-3 carbon atoms gives rise to a cluster
of signals usually dominated by the signal for acyl chains
which are saturated or have a double bond well removed from
the methyl end. The minor signals derive from unsaturated
acids in which the double bond is close enough to influence
the shifts of the ω1-3 signals. Apart from the allylic
carbon atoms which have their own characteristic signals,
γ-olefinic carbon atoms are the most significant. Such
compounds usually produce the signals of lowest chemical
shift in each cluster with different shifts for the cis and
trans isomers. Typical results are given in Table 15. For

Table 14   Distribution of EPA and DHA between the α and β
           positions in the white muscle of the Atlantic
           salmon[13]

			α (%)	β (%)
Δ5 acids (EPA + AA)	C2 33.55 (β)	33.36 (α)	57.0	43.0
Δ4 acids (DHA)	C1 172.51 (α)	172.11 (β)	18.4	81.6

Table 15   Some chemical shifts (ppm) for ω1-3 carbon atoms
           in partially hydrogenated oils

	normal signal	signal γ to the double bond		isomer
ω1	14.12	13.80	13.66	Δ14 (c, t)
ω2	22.72	22.37	22.21	Δ13 (c, t)
ω3	31.95	31.56	31.43	Δ12 (c, t)

example, the 12c and 12t esters have characteristic ω3
signals. These isomers are of particular interest because
they are likely to be important products in hydrogenated
linoleate.

It is not possible to identify all the signals which
appear in the clusters for cis and trans allylic carbon
atoms around 27.2 and 32.6 ppm respectively but the
integrals for each category can be summed to give a
cis/trans ratio. Absolute levels can only be determined if
the content of saturated esters is determined by some
independent procedure such as gas chromatography. For
extensively hydrogenated oils the cis/trans ratio is
usually close to the equilibrium value of 20:80.

The part of the spectrum relating to olefinic signals
(128-132 ppm) is very complex but repays careful study
since it is possible to identify most of the signals by
comparison with existing information for cis and trans 18:1
acids.[3] Table 16 contains a list of the olefinic signals
observed in 12 hydrogenated fats and their assignment.
Further details are given in the full paper.[16]

Triacylglycerols and phospholipids

Triacylglycerols, phosphatidylcholines, and
phosphatidylethanolamines can be distinguished through
signals associated with the carbon atoms of the head group
and with the acyl carbon atoms (C1). A mixture of
triacylglycerols, phosphatidylcholines, and
phosphatidylethanolamines, each obtained from soybean oil,
was examined. The signals which are distinctive for each

Table 16   Chemical shifts (ppm) of olefinic signals
           observed in 12 hydrogenated fats

double bond position	trans		cis	
5	131.85	128.64	–	–
6	131.05	129.41	–	–
7	130.74	–	–	–
8,14	130.59	⌈130.09 ⌊130.04	⌈130.09 ⌊130.04	129.58
9	130.49	130.18	130.02	129.70
10	130.43	130.25	129.95	129.78
11-13		⌈130.40 ⌊130.30		⌈129.92 ⌊129.83

Table 17   Chemical shifts (ppm) for selected signals observed in a mixture of triacylglycerols, phosphatidylethanolamines, and phosphatidylcholines derived from soybean oil.

	TAG	PE	PC
C1α	173.91	174.11	174.14
C1β	173.49	173.74	173.79
G1	62.46	62.86	62.86
G2	69.38	⌈70.72	⌈70.72
		⌊70.62	⌊70.62
G3	62.46	⌈64.06	63.88
		⌊63.99	
CH$_2$O	–	⌈61.98	⌈59.44
		⌊61.91	⌊59.38
CH$_2$N$^+$	–	40.87	66.65
Me$_3$N$^+$	–	–	54.30

(assignments based on study of individual components)

lipid class are collected together in Table 17. There are six distinct C1 signals - the α and β signals for each type of lipid - and three different signals for C3 of the glycerol unit (G3). There are also characteristic signals for the CH$_2$O, CH$_2$N$^+$, and N$^+$Me$_3$ carbon atoms. Several of the signals are split through coupling with the phosphorus atom. (F.D. Gunstone, unpublished).

Glycerol esters

The ^{13}C nmr spectra of glycerol esters, acetylated monoacylglycerols, and propylene glycol esters provide useful information on the nature of natural and commercial mixtures. These compounds are used as food emulsifiers within the broad range of permitted food additives.

Glycerol can form five different kinds of esters: two monoacylglycerols (1- and 2-), two diacylglycerols (1,2- and 1,3-) and the triacylglycerols. These show nmr signals with different chemical shifts for the three glycerol carbon atoms. The "symmetrical" esters (2-monoacyl, 1,3-diacyl, and triacyl) show only two glycerol signals in a 1:2 intensity ratio but the "unsymmetrical" esters (1-monoacyl and 1,2-diacyl) have three glycerol signals of equal intensity. The chemical shifts for these various carbon atoms in CDCl$_3$ and CDCl$_3$-CD$_3$OD (2:1) solution are listed in Table 18 along with the chemical shifts for glycerol itself. The intensities which accompany the chemical shifts can be used in a semiquantitative manner.[17]

Acetylated monoacylglycerols are more complex mixtures and six different species have been identified from their nmr spectra. These are based on signals for the glycerol carbon atoms, the acyl carbon atoms (C1), and the two

Table 18. Chemical shifts (ppm) for glycerol carbon atoms
in CDCl₃ and CDCl₃ - CD₃OD solution

glycerol ester	CDCl₃[a]	CDCl₃-CD₃OD[b] (2:1)
β-carbon atom		
2-mono-	74.97	75.36
1,2-di	72.25	72.40
1-mono-	70.27	70.24
tri-	68.93	69.41
1,3-di	68.23	67.71
α-carbon atom		
1-mono-	65.04	65.58
1,3-di	65.04	65.36
1-mono-	63.47	63.46
1,2-di-	62.20	62.92
tri	62.12	62.50
2-mono-	62.05	61.06
1,2-di-	61.58	60.82

<u>a</u>   reference (17)
<u>b</u>   F. D. Gunstone, unpublished
Chemical shifts for glycerol in CDCl₃-CD₃OD are 72.78 (β)
and 63.66 (α)

carbon atoms of the acetyl group. The chemical shifts are
given in Table 19 (F. D. Gunstone, unpublished).

     Esters of propylene glycol (propane-1,2-diol) are of
three kinds: the 1- and 2-monoacyl esters and the 1,2-
diacyl esters. Each type has characteristic chemical
shifts for six carbon atoms - three in the propane unit and
three (C1-3) in the acyl chains (Table 20). These can be

Table 19 Chemical   shifts   (ppm)   of   acetylated
          monoacylglycerols in CDCl₃

	1	2	3	4	5	6
G1	65.15	62.07	65.00	62.00	62.46	62.35
G2	70.26	72.39	68.14	69.16	72.03	68.76
G3	63.39	61.40	65.26	62.33	61.32	62.35
C1	174.31	173.84	173.96	173.38	173.52	173.01
COCH₃ (β) ⌈	-	21.00	-	20.88	-	-
⌊	-	170.98	-	170.16	-	-
COCH₃ (α) ⌈	-	-	20.79	20.69	20.74	20.69
⌊	-	-	171.10	170.57	170.65	170.57

	α	β	α	
1	⌈ OCOR	OH		OH
2	⌈ OCOR	OAc		OH
3	⌈ OCOR	OH		OAc
4	⌈ OCOR	OAc		OAc
5	⌈ OAc	OCOR		OH
6	⌈ OAc	OCOR		OAc

<u>Table 20</u>   Chemical   shifts   (ppm)   for   propylene   glycol
             esters in CDCl₃.

	P1	P2	P3	C1	C2	C3
1-monoacyl	69.46	66.13	19.19	173.97	34.23	24.97
2-monoacyl	65.92	71.77	16.25	173.97	34.58	25.03
1,2-diacyl	65.92	67.98	16.54	⌈173.51	⌈34.51	⌈24.97
				⌊173.24	⌊34.19	⌊25.03

P1-3 refer to carbon atoms in the propane unit, C1-3 refer
to carbon atoms in acyl chain(s).

used along with their intensities to determine the
composition of propylene glycol ester mixtures   (F. D.
Gunstone, unpublished).

   These examples of high resolution ¹³C nmr spectroscopy
show how this technique can be usefully applied to problems
of lipid analysis.

References

1. F. D. Gunstone, M. R. Pollard, C. M. Scrimgeour, N. W.
    Gilman, and B. C. Holland, <u>Chem</u>. <u>Phys</u>. <u>Lipids</u>, 1976,
    <u>17</u>, 1.

2. J. Bus, I. Sies, and M. S. F. Lie Ken Jie, <u>Chem. Phys.
    Lipids</u>, 1976, <u>17</u>, 501.

3. F. D. Gunstone, M. R. Pollard, C. M. Scrimgeour, and
    H. S. Vedanayagam, <u>Chem. Phys. Lipids</u>, 1977, <u>18</u>, 115.

4. J. Bus, I. Sies, and M. S. F. Lie Ken Jie, <u>Chem. Phys.
    Lipids</u>, 1977, <u>18</u>, 130.

5. F. D. Gunstone, <u>Advances in Lipid Methodology-Two</u> (ed.
    W. W. Christie), The Oily Press, Dundee, 1993, Chapter
    1, 1-68.

6. M. S. F. Lie Ken Jie, C. C. Lam, and B. F. Y. Yan, <u>J.
    Chem. Research</u> (<u>S</u>) 1992, 12 and (<u>M</u>) 1992, 250.

7. F. D. Gunstone, <u>J. Amer. Oil. Chem. Soc.</u>, 1993, <u>70</u>,
    361.

8. F. D. Gunstone, <u>Chem. Phys. Lipids</u>, 1993, <u>65</u>, 155.

9. E. Bascetta and F. D. Gunstone, <u>Chem. Phys. Lipids</u>,
    1985, <u>36</u>, 253.

10. F. D. Gunstone, <u>J. Amer. Oil Chem. Soc.</u>, (1993) <u>70</u>,
    1139.

11. F. D. Gunstone, <u>Chem. Phys. Lipids</u>, 1991, <u>59</u>, 83.

12. M. Aursand and H. Grasdalen, <u>Chem.</u> <u>Phys.</u> <u>Lipids</u>, 1992, <u>62</u>, 239.

13. M. Aursand, J. R. Rainuzzo, and H. Grasdalen, <u>J.</u> <u>Amer.</u> <u>Oil</u> <u>Chem.</u> <u>Soc.</u>, submitted for publication.

14. F. D. Gunstone, <u>Chem.</u> <u>Phys.</u> <u>Lipids</u>, 1991, <u>58</u>, 159.

15. F. D. Gunstone, <u>Chem.</u> <u>Phys.</u> <u>Lipids</u>, 1990, <u>56</u>, 201.

16. F. D. Gunstone, <u>J.</u> <u>Amer.</u> <u>Oil</u> <u>Chem.</u> <u>Soc.</u>, 1993 <u>70</u>, 965.

17. F. D. Gunstone, <u>Chem.</u> <u>Phys.</u> <u>Lipids</u>, 1991, <u>58</u>, 219.

Application of Modern Mass Spectrometric Techniques to the Analysis of Lipids

R. P. Evershed

DEPARTMENT OF CHEMISTRY, UNIVERSITY OF BRISTOL,
CANTOCK'S CLOSE, BRISTOL BS8 1TS, UK

ABSTRACT

This contribution provides an overview of some of the newer
sample introduction, ionisation and mass analysis techniques
for use in the analysis of lipids. Special consideration is
given to the use of techniques for the study of intact acyl
lipids, i.e. those techniques which complement, or avoid the
need to use, degradative (chemical or enzymatic) methods for
releasing fatty acids from complex lipids such as
acylglycerols. A detailed description is given of the use of
high-temperature gas chromatography/mass spectrometry (HT-
GC/MS) for the analysis of triacylglycerols and steryl fatty
acyl esters. Other techniques that are discussed include:
fast atom bombardment-mass spectrometry (FAB-MS), high
performance liquid chromatography/mass spectrometry (LC/MS),
electrospray (ESP) and atmospheric pressure chemical
ionisation (APCI) techniques for the analysis of high
molecular weight and polar lipids. Consideration is also
given to the use of tandem mass spectrometry (MS/MS) for
mixture analysis and structure investigations.

KEYWORDS: Intact acyl lipids, GC/MS, LC/MS, SFC/MS, tandem
mass spectrometry, MS/MS, mixtures analysis, structure
investigations.

1 INTRODUCTION

A fundamental problem that exists in the analysis of lipids
derives from the structural and compositional complexity of
the mixtures that occur in living organisms. For instance,
triacylglycerols mixtures are often highly complex since the
number of possible positional and stereochemical isomers =
[the number of different fatty acids]^3. Hence, in the case
of a triacylglycerol containing only two different fatty
acids, such as palmitic and stearic acids, eight structures
are possible (Figure 1).

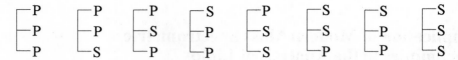

Figure 1. Possible structures for triacylglycerols containing combinations of only two fatty acids, i.e. palmitic (P) and/or stearic (S) acids.

Since naturally occurring triacylglycerol mixtures frequently contain much larger numbers of individual fatty acids, even when certain isomeric forms are ignored on biosynthetic grounds, the natural mixtures may be too complicated to be fully resolved by the currently available analytical techniques despite great advances(see Table 1; [1]).

Table 1. The number of possible triacylglycerol isomers possible when mixtures of 5, 10 and 20 different fatty acids are present (Modified from reference 1).

Isomer consider- ations	Number of triacylglycerols		
	Number of fatty acids		
	5	10	20
All isomers considered	125	1000	8000
Optical isomers excluded	75	550	4200
Isomers not distinguished	35	220	1540

Although the situation is not quite so complicated in the case of simpler lipids, such as wax and steryl fatty acyl esters, their analysis by conventional means can still raise ambiguities. Analytical protocols for such compounds generally comprises steps of: (i) extraction; (ii) fractionation; (iii) chemical or enzymatic hydrolysis; (iv) chemical modification (derivatisation); (v) chromatography, and (vi) spectroscopic analysis. The exact components of the analytical strategy will depend upon the specific analytical goals. Gas chromatography (GC) or combined GC/mass spectrometry (GC/MS) are frequently used to determine the nature of the individual components of the alcohol and fatty acid fractions. Ambiguities arise in this approach due to: (i) the presence of contaminating acyl lipids which have similar chromatographic properties to the targeted class and that will contribute fatty acids upon hydrolysis, and (ii) the mixture of fatty acids and alcohols produced being so complex that it is impossible to deduce the exact composition of the original intact acyl lipid mixture (Figure 1). In spite of apparent shortcomings, degradative methods are routinely used in lipid analysis, since the simpler lipid moieties produced are readily analysed by gas chromatography (GC).

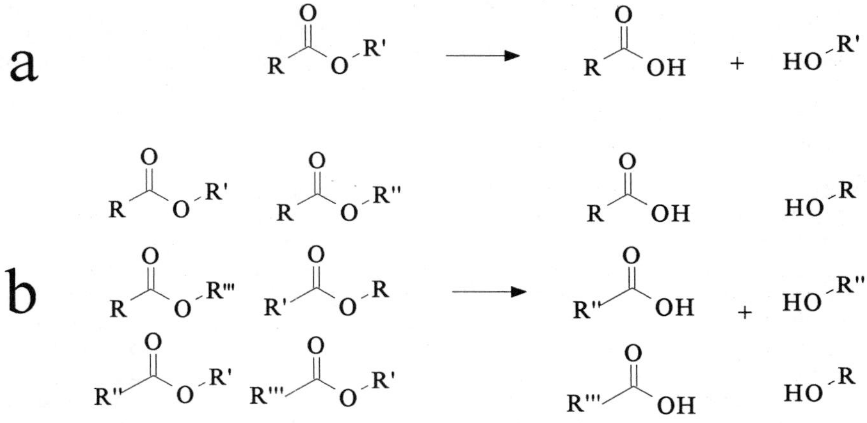

Figure 1. Schematic representation of the loss of compositional integrity resulting through chemical or enzymatic hydrolysis of acyl lipids. (The original composition of the intact lipid is relatively easily deduced when only a single alcohol or fatty acid result (a). However, when a mixture of alcohols and a mixture of fatty acids is produced (b) problems exist in deducing which alcohol was bound to which acid in the intact lipid mixture).

Moreover, the ability to connect flexible fused silica GC columns directly to modern mass spectrometers, without needing to incorporate complicated interface systems, affords a very powerful means of on-line separation and analysis [2-4] and there is an enormous body of published mass spectral information that can assist in making identifications [5-8].Considerable work has also been performed in developing chemical derivatisation techniques for fatty acids and other long-chain alkyl compounds in an effort to provide more detailed structure information, such as the location and geometry of double-bonds in unsaturated compounds, on the basis of data obtained directly from

GC/MS analyses [3, 7, 9 and 10 and references therein].

The recent developments that have occurred in the fields of chromatography and mass spectrometry means that new possibilities exist for the direct analysis of mixtures of intact complex lipids. It is the intention of this article to provide an overview of some of the new mass spectrometric techniques that are available for use in the study of lipids. Mention is made of well-established techniques where appropriate, for the sake of completeness, comparative purposes, and as background for subsequent discussions. Published works which represent more comprehensive treatments of earlier work in this field are to be found in references [5-8]. Specific areas that will be covered in this article include: (i) sample introduction and ionisation techniques that have been used in the analysis of lipids and lipid mixtures (not including combined chromatographic/mass spectrometric methods); (ii) high temperature capillary GC/MS analysis of intact acyl lipids; (iii) supercritical fluid chromatography/MS (SFC/MS) and high performance liquid chromatography/MS (LC/MS) for the separation and on-line mass spectral analysis of lipid mixtures, and (iv) use of tandem mass spectrometry (MS/MS) with low and high collision energy regimes for mixture analysis and structure investigations. Since this is not a comprehensive review the author would like readers to note that the selection of particular examples from the literature does not imply that these are the first or sole applications of a particular technique or approach.

2 IONISATION TECHNIQUES

The earliest examples of the use of mass spectrometry for the analysis of complex or high molecular weight lipids employed the heated direct insertion probe for sample introduction, in conjunction with electron ionisation (EI)or chemical ionisation (CI). Although relatively apolar lipids such as triacylglycerols, steryl fatty acyl esters and wax esters yield some structurally diagnostic fragment ions under these conditions detailed compositional information was inevitably limited due to the complexity of the mixtures examined [11-15; see also above]. In addition, more polar compounds, such as phospholipids, are not amenable to analysis by the heated direct probe technique due to their thermal instability. A wide variety of sample introduction and ionisation techniques has been explored with the aim of extending the range of applications of mass spectrometry, particularly in relation to the analysis of the more polar and high molecular weight compounds. Significantly, many of these new techniques have also been found to be of value in the analysis of apolar compounds. Some of the general trends in the mass spectrometric behaviour of various classes of lipid under a variety of sample introduction and ionisation modes are discussed below.

2.1 Electron ionisation (EI)

Apolar lipids such as triacylglycerols, wax esters and steryl fatty acyl esters are readily studied using the direct probe technique for sample introduction. Under EI conditions triacylglycerols fragment readily. An $M^+\cdot$ ion is generally observed together with an ion due to the loss of water, $[M-18]^+$, but both these ions occur at <1% relative abundance. The most abundant fragment ions seen at high mass correspond to the loss of an acyloxy group from the molecular ion, $[M-RCO_2]^+$ which can be used to determine the nature (but **not** the position, i.e. *sn*-1, *sn*-2 or *sn*-3) of the acyl moieties attached to the glycerol backbone. Further evidence for the nature of the acyl moieties come from the *m/z* values of the acylium ions, $[RCO]^+$, which are seen at lower mass. It has been widely reported that the presence of unsaturation in the acyl moiety affords fragment ions possessing a ketene structure, $[RCO-1]^+$, of greater abundance than the acylium ions referred to above. However, recent high-temperature GC/MS work in our laboratory (discussed in more detail below) has shown that this rule does not always hold true and at an ion source temperature of 300°C acylium ions are seen in greater abundance than the expected ketene-type, $[RCO-1]^+$ ions.

Probe EI-MS analysis of complex triacylglycerol mixtures affords complex spectra in which clusters of ions are seen corresponding to the presence of fatty acyl moieties separated by *m/z* according to their differing carbon numbers and degrees of unsaturation. Mixtures of steryl and wax esters can be surveyed using the direct probe technique in conjunction with EI, although, problems will again be encountered in interpreting the spectra due to the complexity of the mixtures involved. The EI spectra of wax esters are generally more informative than those obtained from steryl fatty acyl ester. In the case of the former an $M^+\cdot$ ion is generally present whereas this is often absent in the case of steryl fatty acyl esters, particularly for those compounds containing a Δ^5 sterol moiety [3]. The CI spectra of steryl fatty acyl esters will be discussed in detail below in considering their analysis by high-temperature GC/MS.

The presence of an $M^+\cdot$ ion in the spectra of wax esters provides an indication of the molecular weight, and hence, carbon number and degree of unsaturation. The principal low mass fragment ion corresponds to the protonated acyl moiety, $[RCO_2H+H]^+$, and is accompanied by an acylium ion of

lower abundance. Ions derived from the alcohol moiety are
much less prominent. The easiest means of deducing the
nature of the alcohol moiety is from the mass difference
between the $M^{+\cdot}$ and $[RCO_2H+H]^+$ ions. Deductions based on
these ions can be corroborated by examining spectra of even
mass alkenyl fragment ions, $[C_nH_{2n}]^+$ [3,5,16].

2.2 Chemical ionisation (CI) and desorption chemical ionisation (DCI)

Where uncertainties surround the interpretations made on
the basis of EI spectra of lipids then the use of CI is
generally considered. Ammonia CI has been found to be of
special value since it affords pseudo- (or quasi-)
molecular ammonium adduct ions, $[M+NH_4]^+$, at relatively low
ion source block temperatures, *ca.* 230°C. These pseudo-
molecular ions are generally some 20-fold more abundant
than the corresponding molecular ions seen when EI is used
[17]. The base peak and sole fragment ion seen in the
positive-ion ammonia CI mass spectra of triacylglycerols
corresponds to $[MH-RCO_2H]^+$. Provided the mixtures under
investigation are not too complex such spectra are of value
in determining the carbon number and degree of unsaturation
of the triacylglycerols and the nature of acyl moieties
they contain. Neither the EI nor CI spectra of
triacylglycerols provide information relating to the
positions of substitution of the acyl groups on the
glycerol backbone, i.e. *sn*-1, *sn*-2 or *sn*-3.

CI-MS has been proposed as a method for the quantitative
assessment of triacylglycerol mixtures. The initial
problems that were encountered which affected precision and
reproducibility were found to be due to fractionation of
mixtures on the probe during 'flash' volatilisation. Such
problems are largely eliminated with modern probe designs,
faster MS scanning and continuous data acquisition. The
availability of desorption chemical ionisation (DCI) offers
a further level of refinement of the direct analysis of
triacylglycerol mixtures. In a recent investigation it was
found that comparable qualitative and quantitative
assessments of the composition of the triacylglycerols
present in the alga *Chlorella kessleri* were achieved using
DCI and GC (following peak identifications by GC/MS) [18].
In this study the relative intensities of the pseudo-
molecular ions were used to assess the percentage
composition of the individual triacylglycerols.

2.3 Field ionisation (FI) and field desorption (FD) mass spectrometry

Other 'soft ionisation' methods that have been applied to
the analysis of triacylglycerols include the related

techniques of field ionisation (FI) and field desorption
(FD) [19]. Although there are few published examples of FI
spectra of triacylglycerols, it appears that the spectra
produced closely resemble those obtained by CI [20]. For
lower molecular weight compounds, $M^{+\cdot}$, $[M-RCO]^+$ and $[RCO]^+$
ions are present, while for higher molecular weight
compounds $M^{+\cdot}$ is absent. In marked contrast to FI and
direct insertion probe techniques (see above), with FD the
sample is ionised without need for prior vaporisation. For
this reason the FD technique is able to provide useful
spectra, containing molecular weight information, from
thermally unstable compounds. FD was the technique of
choice for this type of analysis prior to the development
of FAB-MS (see below). Little or no fragmentation was
observed in the FD mass spectra of triacylglycerols at
optimum emitter currents; $M^{+\cdot}$ and/or $[M+H]^+$ ions were the
base peaks in the spectra of both low and high molecular
weight triacylglycerols. At higher than optimum emitter
currents thermally induced decompositions occurred which
lead to structurally significant fragment ions, e.g. [M-
RCO]$^+$ and [RCO]$^+$ [20-22].

FD has been used to provide molecular weight profiles of
natural oils [22]. The 'soft' ionising property of FD-MS
has also been found to be of value in deducing the nature
of fatty acyl moieties in steryl ester analyses [23,24].
FD-MS produced abundant $M^{+\cdot}$ ions for intact cholesteryl
esters during analyses of crude lipid extracts of human
plasma. Ions were detected at m/z 622, 624, 646, 648, 650
and 672 which corresponded to cholesteryl palmitoleate
(16:1), palmitate (16:0), linolenate (18:3), linoleate
(18:2), oleate (18:1) and arachidonate (20:4) respectively.
The intensity distribution of the $M^{+\cdot}$ ion signals was in
approximate agreement with the established fatty acid
content of human plasma. Since differential ionisation of
the various ester species was found to occur, stable
isotope labelled internal standards would be required for
accurate quantitative analysis by this approach.
Interestingly, the FD response was found to increase with
increase in the number of double bonds in the fatty acyl
moiety.

2.4 Fast atom bombardment mass spectrometry (FAB-MS)

Although generally regarded as a technique for the analysis
of relatively polar or surface active substances fast atom
bombardment-mass spectrometry (FAB-MS) and liquid secondary
ion mass spectrometry (LSIMS) can be used in the analysis
of triacylglycerols [25] with *m*-nitrobenzyl alcohol which
has been found to be a useful matrix. In a recent
investigation [26], neat samples of mono-, di- and
triacylglycerols were analysed by positive and negative-ion
FAB-MS which yielded $[M-RCOO]^+$ and $[RCOO]^-$ ions without

significant pseudo-molecular ions. However, when samples were dissolved in *m*-nitrobenzyl alcohol solution containing NaI, abundant [M+Na]$^+$ species were observed together with the [M-RCOO]$^+$ ions. Collision experiments performed on the cationated pseudo-molecular ions afforded identification of each acidic moiety in the acylglycerol species. The methodology was applied to the analysis of olive and seed oils. The use of tandem mass spectrometry in the analysis of lipids is discussed in detail in a separate section below.

Phospholipids are encountered very commonly in total lipid extracts of biological tissues and fluids. Conventionally, their analysis involves adsorption chromatography, e.g. thin layer chromatography (TLC), followed by chemical degradation, generally transmethylation, then GC and/or GC/MS to confirm the identities of the fatty acyl moieties present in the parent phospholipid [7]. An alternative strategy involves using phospholipase C to release diacylglycerols which can then be analysed by GC and/or GC/MS after derivatisation, e.g. TMS, TBDMS or acetate [27,28], to provide detailed structure information, including the position of substitution of the acyl groups on the glycerol backbone [3,29,30]. The FAB-MS and LC/MS approaches can be of advantage in analyses of phospholipids since they avoid the need for chemical or enzyme treatments. The LC/MS of phospholipids will be discussed in detail below in a section devoted to the LC/MS of lipids.

FAB-MS offers a very useful means of obtaining structure information from intact phospholipids. Both the positive-ion [31,32] and negative-ion [33,34] scanning modes can yield structurally diagnostic ions. Very briefly, positive-ion FAB-MS yields spectra which contain protonated and cationated pseudo-molecular ions. Fragment ions are seen corresponding to mono- and diacylglyceryl, and phosphate ester moieties. Negative ion spectra display ions of low abundance which correspond to deprotonated molecular anions, [M-H]$^-$, the loss of single fatty acyl moieties and the phosphate ester head group. The most prominent ions in the negative ion spectra usually correspond to carboxylate anions, [RCOO]$^-$, usually as the base peak, which are of obvious value in deducing the carbon number and degree of unsaturation of the fatty acyl moieties present in the parent phospholipid.

Problems that can arise in the analysis of phospholipid mixtures were exemplified by a report describing the analysis of egg yolk lecithins (L-α-phosphatidylcholines). The presence of the established combinations of fatty acyl moieties was readily confirmed on the basis of the masses of the [M+H]$^+$ ions recorded in the positive-ion scanning mode. However, the facile expulsion of two hydrogens (presumed to be from the glyceryl chain) could lead to erroneous conclusions concerning the presence of more highly unsaturated acyl moieties [31]. Further complications can arise in the negative ion spectra of phospholipids since the highest mass ion seen in phosphatidylcholines corresponds to $[M-CH_3]^-$ which is isobaric with the highest mass ion, $[M-H]^-$, of a phosphatidylethanolamine bearing the same fatty acyl moieties. This ambiguity can be resolved either by comparing the chromatographic mobilities (TLC or HPLC) of the respective phospholipid fractions or by employing positive-ion FAB-MS, since both the cholines and ethanolamines display $[M+H]^+$ ions.

2.5 Plasma desorption mass spectrometry (PDMS)

Californium-252 plasma desorption can yield diagnostic spectra from acyl lipids. This was revealed in an investigation aimed at discovering the source of contaminating ions arising from careless handling of sample holders [35]. The principal source of the contamination was found to be triacylglycerol originating from finger lipids. This was deduced by recording the PD mass spectra of representative members of the major classes of known skin lipids, including a fatty acid, methyl and wax ester, mono-, di- and triacylglycerols, in addition to an anhydride. Similar trends in fragmentation and ionisation were noted in all the compounds studied. The positive ion PD spectrum of fingerprint lipids is shown in Figure 2.

**SKIN LIPIDS
(POSITIVE IONS)**

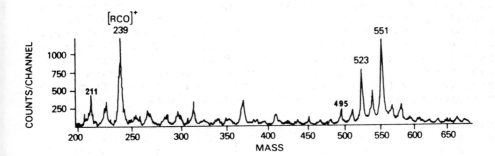

Figure 2. Positive ion plasma desorption mass spectrum of fingerprint lipids. The cluster of ions in the m/z range 470 to 600 correspond to $[M+H-RCOOH]^+$ ions of triacylglycerols. Reproduced with permission of John Wiley and Sons Ltd. from reference 35.

The PD mass spectrum of authentic tripalmitin contained a
weak [M+H]$^+$ ion. The base peak in the spectrum was found to
be [M+H-RCOOH]$^+$, appearing at m/z 551, while at lower mass
a prominent acylium ion, [RCO]$^+$ (m/z 239) was clearly
visible. All these latter ions are evident in the
fingerprint spectrum shown in Figure 2. The production of
higher mass ions via attachment processes was seemingly a
common event. For example, an ion at m/z 1045 in the
spectrum of tripalmitin was presumed to arise by attachment
of an acylium ion to a neutral tripalmitin species. The
pattern of ions observed for tripalmitin confirmed that the
triacylglycerols present in skin lipids were the principal
source of contaminating ions in PDMS analyses. Negative ion
scanning produced spectra that were dominated by
carboxylate anions.

2.6 Electrospray ionisation mass spectrometry (ESP-MS)

As with FAB-MS and PD-MS, electrospray ionisation is not
generally regarded as a mass spectrometric technique that
would be considered for routine use in the analysis of
lipids. However, there is an increasing body of evidence to
suggest that most classes of lipid are amenable to ESP-MS
analysis. For example, positive-ion electrospray spectra
have been recorded for mixtures of acylglycerols using a
carrier solvent of chloroform:methanol (70:30 v/v) modified
with either alkali metal ion, ammonium salts, or formic
acid, in order to promote cationation of the sample
molecules [36]. Positive ion current signals for [M+Na]$^+$
and [M+NH$_4$]$^+$ ions were observed for acylglcerols introduced
at low concentrations with no fragmentation being
detectable. There was found to be a significant decrease in
the ion current as the polarity of the analytes decreased.
Hence, acylglycerols containing unsaturated fatty acyl
moieties exhibited a greater response compared to compounds
containing fully saturated acyl moieties, while the ion
currents generated by equimolar amounts of monoacylglycerol
were greater than those produced by diacylglycerols which
in turn were greater than those for triacylglycerols. The
use of ESP ionisation in conjunction with tandem mass
spectrometry for generating structure information from
mixtures of acyl lipids is discussed in detail below.

ESP-MS has also been found to be of use in generating
structure information from steryl fatty acyl esters and
phospholipids. In the case of phosphatidylcholine positive
ion electrospray ionisation yielded largely only molecular
weight information when samples were run in a carrier
medium comprising isopropanol:water (50:50 v/v) with and
without added ammonia solution (0.1% v/v). As can be seen
from the results presented in Figure 3 the addition of

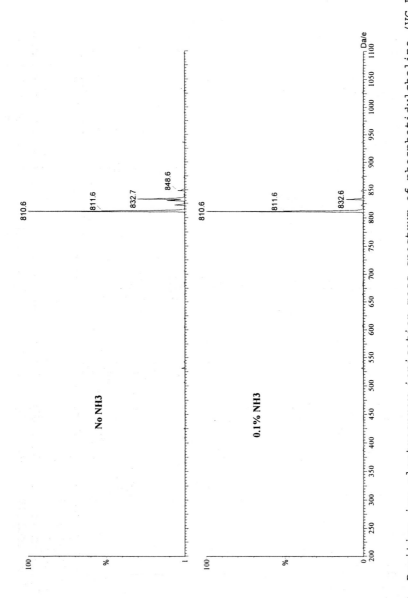

Figure 3. Positive ion electrospray ionisation mass spectrum of phosphatidylcholine (VG Biotech U.K. are acknowledged for permission to present these data).

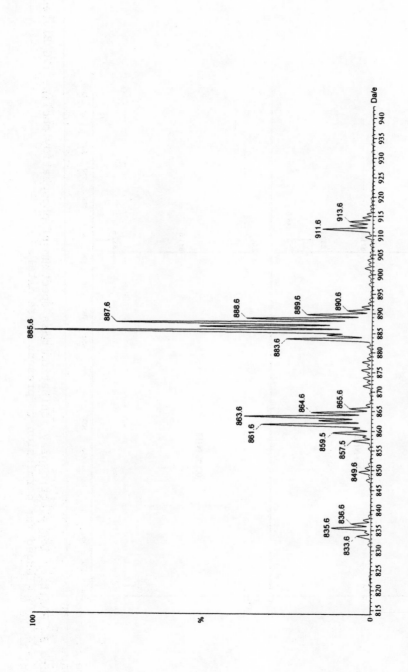

Figure 4. Pseudo-molecular ion region of the negative ion electrospray mass spectrum of phosphatidylinositol (VG Biotech U.K. are acknowledged for permission to reproduce these data).

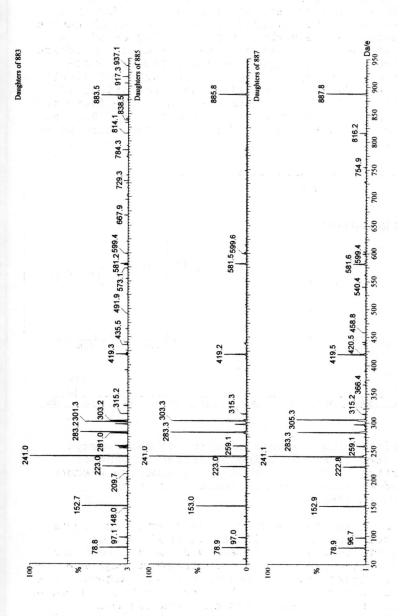

Figure 5. Product ion spectra of [M-H]⁻ ions, m/z 883, 885 and 887, shown in Figure 4. The presence of an [RCOO]⁻ ion at m/z 283 confirms the presence of a C₁₈:₀ moiety in each compound. The ions at m/z 301, 303 and 305 confirm the presence of C₂₀:₅, C₂₀:₄ and C₂₀:₃ acyl moieties is respectively in the three targeted compounds (VG Biotech U.K. are acknowledged for permission to present these data).

ammonia would appear to suppress the sodium adduct ion and promote molecular ion formation. The favourable positive ion behaviour is presumably enhanced by the presence of a quaternary ammonium moiety. The negative spectrum of PC was dominated by [RCOO]⁻ ions at low mass with [M-2H]⁻ and [M+HCOO-H]⁻ ions being present at high mass. Product ion spectra of the latter ion yielded [RCOO]⁻ ions which can be used to characterise the carbon number and degree of unsaturation of the acyl moieties. The molecular ion region of the negative ion electrospray mass spectrum of phos-phatidylinositol is shown in Figure 4. The array of [M-H]⁻ ions that is seen derives from different combinations of carbon numbers and degrees of unsaturation in the fatty acyl moieties. The product ion spectra for the *m/z* 883, 885 and 887 ions are shown in Figure 5 and confirm the nature of the fatty acyl moieties.

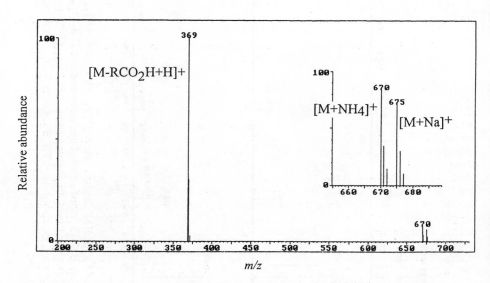

Figure 6. Positive-ion ESP mass spectrum of cholesteryl stearate analysed by ESP-MS using a carrier solvent of chloroform/methanol (2:1 v/v) containing ammonium acetate. The sodium adduct arises from residual ions in the analytical system [38].

Steryl fatty acyl esters analysed in an exactly analogous fashion have been found to yield, ammoniated pseudo-molecular adduct ions, [M+NH₄]⁺, and a major fragment ion corresponding to the protonated steradienyl moiety, i.e. [M-RCO₂H+H]⁺ (see Figure 6). Although no ions characteristic of the fatty acyl moiety are seen, the carbon number and degree of unsaturation can be readily deduced from the mass difference between the pseudo-molecular adduct ion and the protonated steradienyl fragment ion. The ESP-MS behaviour of

steryl esters under these conditions is largely analogous to
that seen when using conventional positive-ion ammonia CI
[39].

3 COMBINED CHROMATOGRAPHIC/MASS SPECTROMETRIC
TECHNIQUES FOR THE ANALYSIS OF LIPID MIXTURES

Gas chromatography is the most widely used technique for the
analysis of lipids and has a role to play in the analysis of
all the major lipid classes. The combination of GC with MS
remains the most effective technique for the analysis of
mixtures of volatile or semi-volatile organic compounds. The
identities of fatty acids released from complex lipids are
readily confirmed by means of GC/MS following derivatisation
(methylation or silylation). More exotic derivatives and
chemical modifications (e.g. pyrolidinyl, dimethyl
disulphide, picolinyl, deuteriation, etc.) can assist in
making very precise structure assignments by means of GC/MS.

3.1 High temperature gas chromatography/mass
spectrometry (HT-GC/MS)

3.1.1 Practical considerations

The analysis of intact acyl lipids such as triacylglycerols,
steryl fatty acyl esters and wax esters, constitutes a very
demanding application of GC/MS. Complex lipids such as
phospholipids are not readily amenable to GC/MS due to their
low volatility and thermal instability even after
derivatisation. However, there is at least one report of the
use in the analysis of the trimethylsilyl derivative of 1-*O*-
hexadecyl(16:0)- and 1-*O*-octadecyl(18:0)-2-lyso-*sn*-glycero-
3-phosphate using a short (1m x 1 mm i.d.) column packed
with 1.5% OV-1 on Chromosorb Q (80-100 mesh)[40]. The
retention times of these two compounds at a GC oven
temperature of 290°C were 1.9 and 2.9 minutes respectively.
No appreciable thermal degradation was seen and EI and CI
spectra were obtained in the GC/MS mode.

Substantial work was carried out in the 1960s and 1970s
using short packed columns with low stationary phase
loadings for the analysis of mono-, di- and triacylglycerols, wax esters and steryl fatty acyl esters and even
acylated steryl glycosides [41-44]. Although analyses of
these relatively involatile substances could be carried out
quite routinely by GC, GC/MS was more problematical due to
the problems of linking packed GC columns to mass
spectrometers at that time. The interface systems available
for packed column GC/MS work were simply unsuitable for
reliable analysis of intact acyl lipids. It was not until
the development of capillary GC columns, particularly the
flexible fused-silica capillaries, that the routine analysis

of intact acyl lipids became possible. There is now a wide
range of columns available that can be used for the GC/MS
analysis of intact acyl lipids such as triacylglycerols,
steryl fatty acyl esters and wax esters. The advantage of
using flexible fused silica capillary columns derives from
the ease of achieving a zero dead volume and chemically
inert connections between the GC and the MS. The optimum
arrangement is to feed the capillary column directly through
the interface line and into the MS ion source. Then,
provided that the transfer line temperature between the GC
and the MS is maintained at around the elution temperatures
of the analytes then high temperature GC/MS analyses can be
performed. Poor high-temperature continuity across the
transfer line is manifested as peak broadening, or more
seriously, complete loss of chromatographic peaks from total
ion chromatograms.

3.1.2 HT-GC/MS using apolar stationary phases

Relatively short (10-15 m) fused silica capillary columns
coated with thin (typically 0.1 μm) films of immobilised
stationary phase such dimethyl polysiloxane have been most
widely used for this type of analysis. Separations of
analytes are largely according to carbon number, with only
partial resolution of molecular species according to their
degree of unsaturation. An excellent review of the HT-GC of
acyl lipids has been given by Mares [45]. The same types of
apolar columns have been found to be useful for the analysis
of di- and triacylglycerols, steryl fatty acyl esters and
wax esters. Figure 7 shows the results of the HT-GC/MS
analysis of the hexane soluble fractions of palm oil
performed in our laboratory. The carbon number resolution
achieved is typical of that obtained when using a cross-
linked dimethyl polysiloxane stationary phase for the
analysis of high molecular weight intact triacylglycerols.
The resolution achieved here is slightly inferior to that
which can be attained when using GC alone due primarily to
the slow scan speed (total scan cycle time *ca.* 3s) of the
mass spectrometer that was used in this analysis. The early
eluting peak corresponds to a single molecular species,
namely tripalmitin, while the later eluting peaks in the
total ion chromatogram (TIC) are mixtures of co-eluting
components as evidenced by the broadening of the peaks in
the TIC compared to the tripalmitin component.

As can be seen from the experimental details given in Figure
7 negative ion chemical ionisation (NICI) with ammonia as
reagent gas was used for this analysis. This ionisation
technique has proved especially valuable in the HT-GC/MS
analysis of triacylglycerols and steryl fatty acyl esters
since structure information is often lacking in EI spectra
due to the unfavourable fragmentation behaviour of these

substances (see above), compounded by interferences from column bleed, which can also prevent satisfactory interpretation of spectra [46]. Figure 8 shows the ammonia NICI mass spectrum of the earliest eluting triacylglycerol peak in the analysis shown in Figure 7. The spectrum is very simple and typical of the type obtained for triacylglycerols under the stated conditions [47]. As indicated the spectrum is dominated by m/z 255 ($[RCO_2]^-$), m/z 237 ($[RCO_2-H_2O]^-$) and m/z 236 ($[RCO_2-H_2O-H]^-$) ions. Although no molecular ion is present the carbon number of the triacylglycerols can be readily assigned on the basis of retention time data for compounds of known structure.

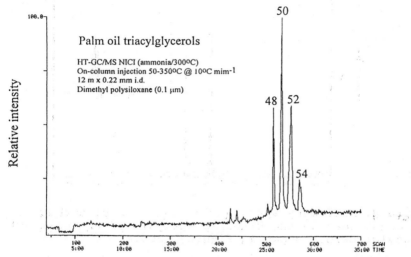

Figure 7. HT-GC/MS analysis of palm oil triacylglycerols. The GC/MS operating conditions are given in the inset to the figure. The numbers on the peaks in the TIC correspond to the total number of carbon atoms associated with the acyl moieties.

Further evidence for the later eluting peaks in the TIC containing mixtures of co-eluting components bearing differing degrees of unsaturation is revealed by examination of their NICI mass spectra. Figure 9 shows the mass spectrum of the last eluting peak (the 54 acyl carbon peak). While the presence of ions in the spectrum at m/z 281, 263 and 262 indicates that the major component of this peak is triolein (53:3), as indicated, the ion at m/z 279 confirms that a significant proportion of the triacylglycerols of the same carbon number bear $C_{18:2}$ moieties. These findings are entirely consistent with the known composition of the triacylglycerols of palm oil. These data serve to illustrate the fact that apolar stationary phases of the type used in this analysis are unable to resolve triacylglycerol species of the same carbon number but which contain acyl moieties

bearing different degrees of unsaturation. The ions observed
in the NICI mass spectra of triacylglycerols appear to be
formed via a gas phase ammonolysis reaction and in effect
provide an on-line fatty acid analysis of the eluting
triacylglycerol peaks in the TIC. The NICI techninque was
originally developed for the analysis of intact steryl fatty
acyl esters in an effort to overcome problems encountered in
EI spectra where structure information concerning the nature
of the fatty acyl moiety was completely lacking.

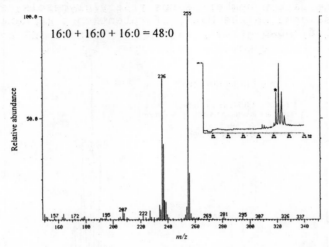

Figure 8. NICI mass spectrum for the 48 acyl carbon
component seen in the HT-GC/MS analysis of the palm oil
triacylglycerols shown in Figure 7 above.

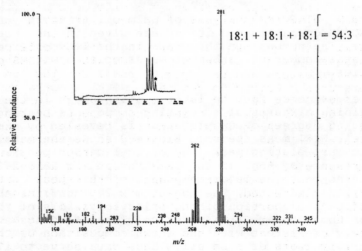

Figure 9. NICI mass spectrum of the 54 acyl carbon peak from
the HT-GC/MS analysis of triacylglycerols shown in Figure 7.

This was especially problematical in the case of esters containing the most commonly occurring sterols, i.e. the Δ^5 compounds. In addition to the fatty acyl derived fragment ions referred to above, the NICI spectra of Δ^5 steryl fatty acyl esters, exhibit deprotonated steradienyl anions which provide information concerning the carbon number and degree of unsaturation of the sterol nucleus in the intact ester (see Figures 10 and 11) [48-54]. The ammonia NICI spectra closely resemble those observed by Field and co-workers in their studies of hydroxyl ion NICI of steroids and their derivatives [55]. Clear parallels can be drawn between the mechanisms of chemical ionisation when either OH$^-$ and NH$_2$$^-$ are the reactive species present in a CI plasma.

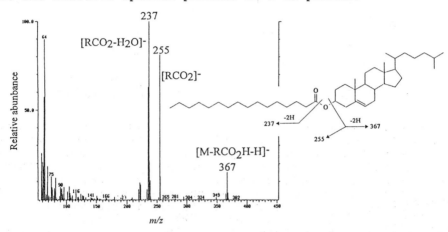

Figure 10. NICI mass spectrum of cholesteryl palmitate recorded by HT-GC/MS at an ion source temperature of 300°C.

The NICI technique has been found to offer significant advantages over positive-ion CI (PICI) in HT-GC/MS of steryl fatty acyl esters and triacylglycerols. The low ion source temperatures required to generate useful spectra in the PICI mode [39,56] can cause cold-trapping of high molecular weight lipids in the short section of column protruding into the ion source. Ammonia NICI spectra of the type shown above in Figures 8-11 are readily obtained at a source temperature of 300°C, and at this temperature cold trapping phenomena are not in evidence [47,48]. HT-GC/MS using ammonia NICI has been shown to of value in the analysis of steryl fatty acyl esters in a wide range of organisms, including higher plants [50,54,57], algae [58], animals [48,49,51,54] and yeast [58]. The technique is of particular value where a complex mixtures of sterols is found esterified with a complex mixture of fatty acyl moieties [57,58]. In these situations

the use of HT-GC/MS in conjunction with both EI and ammonia
NICI is the preferred approach.

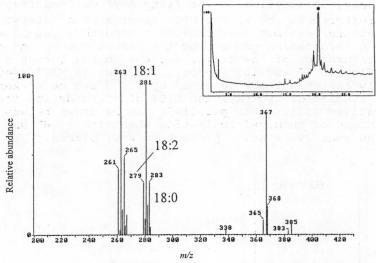

Figure 11. NICI mass spectrum of the peak asterisked in the
inset TIC obtained by HT-GC/MS of the steryl ester extract
of the tick *Rhipacephalin appendiculatus*. The ion at *m/z* 367
corresponds to the cholestadienyl anion and confirms that
only cholesteryl esters are present. The [RCO$_2$]$^-$ ions at *m/z*
283, 281 and 279 indicate that the peak comprises co-eluting
cholesteryl 18:1, 18:2 and 18:3 esters in reasonable
abundance. At longer and shorter retention times then the
asterisked peak cholesteryl esters bearing saturated fatty
acyl moieties with 14, 16, 20 and 22 carbons atoms were also
detected.

3.1.3 HT-GC/MS using polarisable stationary phases

An important advance in the GC and GC/MS analysis of intact
acyl lipids, such as triacylglycerols, has been the
development of high-temperature stable polarisable
immobilised stationary phases such as OV-22 (65% phenyl
methyl polysiloxane) which is thermally stable up to 360°C.
A notable property of this phase is its ability to separate
molecular species that exhibit only very small differences
in polarity, i.e. triacylglycerols possessing different
numbers of double bonds. While the effectiveness of this
phase for the GC analysis of triacylglycerols has been
demonstrated in our own work (unpublished) and in a number
of published papers, e.g. [59,60], there are few reports of
its use in applications involving HT-GC/MS. Kuksis and co-
workers have provided an elegant example of the use of this
phase in the GC and GC/MS analysis of a volatile butter oil
(lower molecular triacylglycerol components) fraction of

bovine milk fat [61]. As already mentioned above one problem that exists in the HT-GC/MS analysis of higher molecular weight triacylglycerols comes from the chemical background in the ion source due to the column bleed into the mass spectrometer. A partial solution to this has been achieved by carrying out HT-GC/MS analyses in the selected ion monitoring mode [62]. Identifications of the nature of the fatty acyl moieties present in individual molecular species rely on matching the retention times (or scan numbers) of different peaks in chromatograms characteristic of particular fatty acyl moieties. The ions that were used to characterise fatty acyl moieties include $[RCO]^+$ and $[M-OCOR]^+$. However, since it is known that at relatively low ion source temperatures, *ca.* 200°C, triacylglycerols containing unsaturated fatty acyl moieties yield more abundant $[RCO-1]^+$ and $[RCO-2]^+$ ions rather than $[RCO]^+$, erroneous deductions may be made, particularly when polyunsaturated moieties are present [see also above and references 3, 46 and 63]. The utility of ammonia NICI for the HT-GC/MS analysis of triacylglycerols on an immobilised 65% phenyl methyl silicone coated capillary has recently been assessed [4,63]. Figure 12 shows the TIC chromatogram obtained for the total bovine butter fat triacylglycerols.

The chromatographic resolution achieved by using this type of phase displays a marked improvement over that which can be achieved using apolar dimethyl polysiloxane phases. The inset to Figure 12 shows that separation of the individual compounds according to the degree of unsaturation of their fatty acyl moieties is readily achieved in HT-GC/MS. The peak assignments were confirmed through their summed NICI mass spectra [see reference 4]. The *m/z* 255, 281 and 283 mass chromatograms show the distributions of the individual $C_{16:0}$, $C_{18:1}$ and $C_{18:0}$ fatty acyl moieties in the triacylglycerols eluting in this part of the chromatogram. The chomatographic resolution shown in this analysis is somewhat inferior to that which can be obtained from this type of column in stand-alone GC analyses. However, the overall performance is impressive, especially when account is taken of the slow scan speed of the mass spectrometer used and given that He, rather than H_2, was used as the carrier gas.

We have recently succeeded in greatly improving chromatographic resolution by employing the Biller-Biemann enhancement technique to produce a mass-resolved total ion chromatogram (Figure 13) from which high resolution background-subtracted mass chromatograms and mass spectra can be generated very conveniently [63]. The data generated is being used to further investigate the structural features of complex triacylglycerols that influence GC elution orders on polarisable stationary phases.

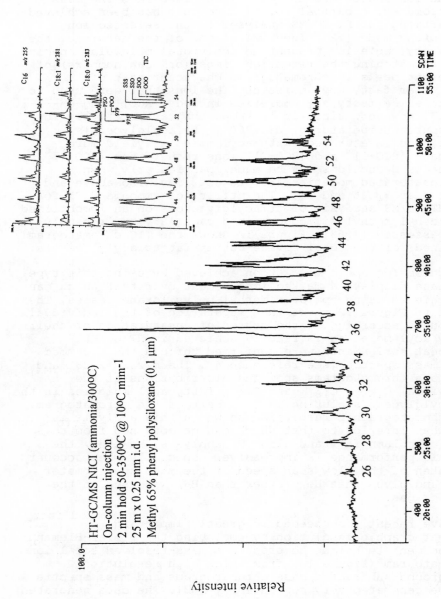

Figure 12. TIC for the triacylglycerols of butter fat using an immobilised 65% phenyl methyl siloxane coated capillary column in conjunction with ammonia NICI. The righthand inset shows the GC/MS operating parameters while the lefthand inset shows the mass chromatograms for the later eluting peaks in the TIC.

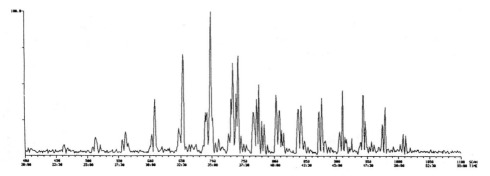

Figure 13. Biller-Biemann mass-resolved chomatogram obtained for the HT-GC/MS analysis of butter triacylglycerols shown in Figure 12.

3.1.4 Consideration of losses of high molecular weight acyl lipids during HT-GC and HT-GC/MS

Although, HT-GC and HT-GC/MS provide very effective means of separating mixtures of triacylglycerols it is important to be aware of the losses of compounds that can occur, particularly of some thermally labile components. Loss of fully saturated species has also been observed and is attributed to irreversible saturation of the stationary phase [45]. Where this is anticipated, such as in analyses of high molecular weight lipids or those containing a preponderance of polyunsaturated fatty acyl moieties, then losses should be assessed by prior analysis of authentic compounds of similar carbon number and degree of unsaturation. In HT-GC/MS analyses this practice also helps in assessing the effectiveness of interface performance, column positioning in the ion source, and MS performance. A useful test for the loss of polyunsaturated components during GC or GC/MS analyses of unknowns involves performing a catalytic reduction and repeating the HT-GC or HT-GC/MS analysis. Significant losses of polyunsaturated components are revealed by the appearance of new peaks in the GC profile or total ion current chromatograms. An example of the use of this technique is demonstrated in Figure 14.

The loss of apolar lipids containing polyunsaturated fatty acyl moieties relates largely to their low thermal stability. Hence, HT-GC/MS is not the method of choice for their routine analysis. Where the presence of lipids bearing polyunsaturated components is suspected then HT-GC/MS analysis must be complemented by analysis of their fatty acids following chemical or enzymatic cleavage, or by catalytic reduction of the intact lipids as discussed above. Where available SFC/MS, HPLC/MS or tandem mass spectrometry (all three of these techniques are discussed in more detail below) now constitute very realistic practical alternatives to HT-GC/MS for the analysis of mixtures of high molecular weight thermally unstable lipids.

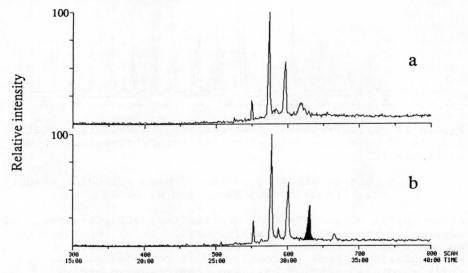

Figure 14. Total ion current chromatogram for the HT GC/MS analysis of a steryl fatty acyl ester extract of the prawn *Penaeus monodon* (a) before and (b) after catalytic reduction [54]. The appearance of additional peaks at long retention time after reduction was due to the presence of steryl fatty acyl esters containing polyunsaturated fatty acyl moieties (20:4, 20:5 and 22:6) that were thermally decomposing during the HT-GC/MS analysis shown in (a), prior the reduction.

3.2 High performance liquid chromatography/mass spectrometry (HPLC/MS)

Although HPLC columns offer substantially less separatory power in terms of the number of theoretical plates they are able to generate, compared to capillary GC columns they offer substantially greater selectivity owing to the possibility for varying both the column packing material and eluent composition between solvents of widely varying polarity, and through the addition of buffers and ion-pairing reagents. The possibility for using HPLC in combination with MS to provide a universal detector is especially important in the case of certain lipids since strong chromophores are frequently lacking. However, a major problem in performing routine LC/MS is the need to preferentially remove a large volume of solvent or vapour prior to introduction of the analyte(s) into the mass spectrometer. The earliest systems used transport devices in

which the HPLC eluent was deposited on a belt [64-66] and
the solvent evaporated prior to the analytes being thermally
volatilised under EI or CI conditions. A perforated metal
belt system was used for the analysis of a wide range of
lipid species [65]. The perforations ensured the effective
entrainment of HPLC eluant at a flow rate of 1 ml min^{-1} and
at a belt speed of 1.8 cm sec^{-1}. With a normal phase silica
HPLC system the eluent applied to the belt was evaporated in
a stream of nitrogen and the analytes transported as a
residue on the belt into a heated reactor where they were
volatilised into a carrier gas stream. When hydrogen was
used as the carrier gas hydrogenolysis and reductive
cracking reactions occurred readily, catalysed by the nickel
present in the stainless steel belt. Approximately 15% of
the total flow was emitted into the mass spectrometer. A
wide range of different types of lipid were tested using
this device, including, triacylglycerols, sterols, steryl
fatty acyl esters, glyceryl ethers, glyceryl ether diesters,
glycerophosphatides, sphingolipids, prostaglandins and fatty
acid methyl esters. The method was also applied to the
analysis of methyl ester ozonides to demonstrate the use of
LC/MS for the determination of double bond positions in
unsaturated fatty acids.

The perforated metal belt was rapidly superseded by belts
made of Kapton TM (Dupont polyimide) [66] which although
reducing metal-catalysed thermolysis reactions still
presented problems when thermally unstable or involatile
lipids were volatilised into the ion source. A recent report
describes the analysis of triacylglycerols from rat adipose
tissue using both LC/MS, employing a belt interface, and
HPLC followed by fraction collection, transmethylation and
GC analysis of the resulting fatty acid methyl esters [67].
The triacylglycerols were identified by using a combination
of their fatty acid content, HPLC retention time and by
comparing EI spectra for the intact triacylglycerols with a
library of EI mass spectra. The agreement between the GC and
LC/MS data was good for most triacylglycerols except that
interpretations of the earlier eluting components was
complicated by poor chromatographic resolution.

The most extensive studies of the LC/MS analysis of intact
acyl lipids have been performed by Kuksis and co-workers who
have devoted their efforts to exploring the use of a direct
liquid inlet based on the Baldwin-McLafferty split CI
design [70-72]. The version used by Kuksis and colleagues
comprises a pinhole orifice (2-5 μm in diameter) positioned
immediately adjacent to the ion source of a quadrupole mass
spectrometer. The total HPLC eluent flows past the orifice
during which *ca.* 1% of the total flow (1-2 ml min^{-1}) is
sampled into the ion source. Ionisation of the dissolved
analytes is achieved by solvent mediated CI. Obviously not
all organic solvents will be suitable for the purposes of

CI. However, solvents which are particularly useful for the reversed phase HPLC of lipids also happen to be excellent reagents in mediating solvent-assisted CI, e.g. acetonitrile and propionitrile. The mass spectrometer scans were limited to masses above m/z 200 due to interferences from solvent ions. Excellent separations of a wide range of mixtures of naturally occurring triacylglycerols were obtained using a Supelcosil-18 reversed phase column with a linear gradient of 30-90% propionitrile in acetonitrile. Under these conditions the spectra of triacylglycerols were dominated by abundant $[M+H]^+$ ions. Low abundance ions of $[M+41]^+$ and $[M+55]^+$ correspond to acetonitrile and proprionitrile adducts, respectively. At lower mass than the protonated molecular ions were diacylglycerol-type ions $[MH-RCOOH]^+$ resulting from the random loss of an acid moiety from the intact molecule. At still lower mass, acylium ions, $[RCO]^+$, were observed.

Reversed-phase HPLC is also well-suited to the separation of diacylglycerols. Again Kuksis and co-workers have been the major influence in this area [69], through using the DLI interface to study diacylglycerols derived by Grignard degradation of triacylglycerols and from glycerolphospholipids by phospholipase C treatment. Although generally thought of as a derivatisation strategy for GC and GC/MS it was found to be convenient to prepare *tert*-butyldimethylsilyl (TBDMS) ethers of diacylglycerols prior to LC/MS. The TBDMS ethers yield spectra largely analogous to those of triacylglycerols under solvent-assisted CI conditions. The most prominent ions corresponded to $[M-RCO_2H+H]^+$ and $[M-TBDMSOH+H]^+$.

The benzoate derivatives of diacylglycerols, which are of special interest due to their UV absorbing properties [71], were also found to yield diagnostic solvent-assisted CI spectra in LC/MS analyses. The most prominent ions that were produced corresponded to $[M-RCO_2H+H]^+$ and $[M-benzoic acid+H]^+$. Although acylium ions were absent from the spectra of saturated and mono-unsaturated fatty acyl moieties they were evident in the spectra of more polyunsaturated species, and their relative abundances were found to increase with increasing unsaturation of the individual fatty acyl moieties [69]. The corresponding pentafluorobenzoates exhibited favourable NICI behaviour yielding deprotonated molecular anions which were used to generate the summed mass chromatograms essential in deconvoluting the complex natural mixtures that are so routinely encountered.

More recently Kuksis and co-workers have explored the use of the 3,5-dinitrophenyl urethane (DNPU) derivatives of diacylglycerols [72]. When 1% of dichloromethane was included in the mobile phase the DNPU derivatives produced chloride attachment spectra, $[M-DNPU+Cl]^-$, which were used to identify and quantify the individual diacylglycerol molecular species (see Figure 15). The negative ion detection of diacylglycerol DNPU derivatives by chloride attachment offered a 100-fold improvement in sensitivity compared to the positive-ion mode. These derivatives were separated on chiral phase columns (containing R(+)-naphthylethylamine polymer chemically bonded to a 300 A wide pore spherical silica). In the positive ion mode the DNPU derivatives yielded prominent ions corresponding to [M-DNPU]$^+$ and [RCO+74]$^+$ but gave no molecular or pseudo-molecular ions.

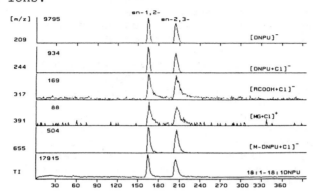

Figure 15. Mass chromatograms for the higher mass dioleylglycerol DNPU in NCI with chloride attachment along with the TIC (bottom). Major ions: m/z 209, [DNPU]$^-$; m/z 244, [DNPU+Cl]$^-$; m/z 317, [RCO$_2$H+Cl]$^-$; m/z 655, [M-DNPU+Cl]$^-$ or [DG+Cl]$^-$; m/z 829, M$^-$. Reproduced from reference 72 with permission of John Wiley and Sons Ltd.

A sensitive assay based on the use of LC/MS with chloride attachment ionisation has been developed by this group for plasma lipid profiling [73]. Negative ion scanning was found to be most effective for the detection of glyceryl esters and ceramides, while positive ion scanning was preferred for the detection of cholesterol and cholesteryl esters. The cholesteryl esters did not yield chloride-attachment spectra in the negative ion mode.

For complex polar lipids the best results have been obtained up to now by means of thermospray ionisation. Kim and Salem [74,75] showed that detailed molecular species analysis could be performed for the most commonly occurring classes of phospholipid. Analyses were conducted in the 'filament on' mode with a high percentage of organic solvent in the carrier solvent. Preliminary investigations with authentic phosphatidylcholines (PC) and phosphatidylethanolamines (PE) showed that pseudo-molecular and simple fragment ions formed readily [74]. The spectra also contained abundant structure information regarding the nature of the fatty acyl moieties and the head groups. Best results were obtained by reversed phase LC/MS using hexane/methanol/0.1M ammonium acetate mixture as mobile phase. The versatility of the method was demonstrated through the analysis of egg phosphatidyl-ethanolamines from which reliable structure information was obtained for each molecular species in *ca.* 15 mins. Further investigation of the technique showed it to be applicable to the analysis of a number of other lipid classes including: phosphatidylinositols, phosphatidylserines, sphingomyelins, triacylglycerols and platelet-activating factor [75]. For these latter classes of lipid, detection limits were found to be in the low microgram range. All the spectra were characterised by the presence of pseudo-molecular ions, such as $[M+H]^+$, $[M+NH_4]^+$ and various cationated species.

Low nanogram detection limits have been reported for ether phosphocholines analysed by thermospray liquid chromatography/mass spectrometry [76]. The amounts and nature of platelet activating factor in human psoriatic skin were determined by using deuterated internal standards for the hexadecyl and octadecyl forms. In a more recent study plasma platelet activating factor levels in the pig were determined using thermospray LC/MS, with baseline values being calculated as 1.5 ng/ml, rising to 8.3 ng/ml following endotoxin administration [77].

Considerable promise is offered by the new ESP and APCI sources for use in LC/MS analyses. The electrospray spectra of phospholipids and cholesteryl esters recorded off-line are shown above in Section 2.6 and there is evidence to suggest that equally useful spectra can be obtained by LC/MS. One group has reported difficulties in the on-line LC/MS analysis of serum steryl esters using an atmospheric pressure ionisation source due to thermal degradation in the nebuliser [78]. It is anticipated that substantial advances will be made in the LC/MS analysis of this, and other important classes of lipids, as these sources become more widely available.

3.3 Supercritical fluid chromatography/mass spectrometry (SFC/MS)

Supercritical fluid chromatography (SFC) has been shown to be of value for the analysis of high molecular weight apolar lipids such as triacylglycerols; both the capillary and packed column variants of the technique have been found to be of use in this area. In one study the triacylglycerols of butter fat were analysed by capillary SFC combined with a double focusing mass spectrometer. Chromatographic separation was obtained according to carbon number using a dimethylpolysiloxane stationary phase (DB-5). The relative proportions of triacylglycerols containing different degrees of unsaturation was deduced by means of selected ion monitoring of molecular ions obtained by EI. Discrimination between fatty acids at the position *sn*-2 and the positions *sn*-1,3 in the triacylglycerol molecules was established by monitoring the $[M-RCO_2CH_2]^+$ ion from reference compounds (see Figure 16) [79].

Brain gangliosides have also been analysed by SFC/MS following derivatisation using both positive ion and negative ion modes [80]. The latter approach was found to offer enhanced sensitivity which was important for quantitative analyses. Abundant pseudo-molecular ion species were obtained from only 20 ng of derivatised gangliosides. The SFC/MS analysis of tocopherols has been achieved using a FAB ion source [81], while the analysis of sterols has been reported using an APCI source [82]. Although SFC/MS offers advantages over GC/MS and LC/MS for the analysis of thermally unstable compounds of low and intermediate polarity it is doubtful that it will ever achieve routine application.

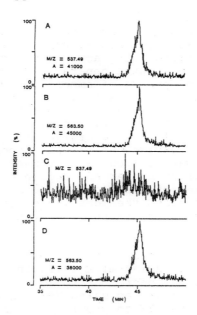

Figure 16. SIM analysis of the $[M-RCO_2CH_2]^+$ ions from the triacylglycerols *rac*-1,2-dihexadecanoyl-3-*Z*-9-octadecenoyl-glycerol (A and B) and 1,3-dihexadecanoyl-2-*Z*-9-octa-decenoyl-*sn*-glycerol (C and D) (Mol. wt. = 832.75). The *m/z* values for the ion of interest are 537.49 for $[M-C_{17}H_{33}CO_2CH_2]^+$ and 563.50 for $[M-C_{15}H_{31}CO_2CH_2]^+$. The chromatographic peaks are normalised to 100% and their peak areas are shown in the Figure. Reproduced from reference 79 with the permission of the American Chemical Society.

4 TANDEM MASS SPECTROMETRY

Tandem mass spectrometry refers to the linking of at least two sequential stages of mass separation. A major advantage of tandem MS is the ability to combine the processes of separation and identification of compounds into a single instrument. An example of the use of tandem MS in the analysis of lipid mixtures has been provided by Henion and co-workers [83]. They used ESP ionisation in conjunction with a quadrupole mass spectrometer for the analysis of mixtures containing mono-, di- and triacylglycerols in a study which provided one of the first examples of the use of ESP in conjunction with non-polar solvents. Compounds were dissolved in a mixture of chloroform and methanol (70:30 v/v) which was modified by the addition of alkali-metal ions or ammonium salts, or by the addition of formic acid, in an effort to encourage the formation of cationated species. Authentic compounds were found to yield $[M+Na]^+$ and $[M+NH_4]^+$ species with picomolar detection limits being readily attainable.

Acylglycerols containing saturated fatty acids were found to give enhanced responses compared to those compounds containing only saturated fatty acyl moieties. Structurally diagnostic fragment ions were generated by MS/MS. Figure 17 shows the product ion spectra of ammoniated monopalmitin, dipalmitin and tripalmitin at a collision energy of 130eV. The most abundant product ions correspond to the loss of fatty acids. Acylium ions were also present in the spectra at low abundance. At collision energies of > 100eV, fragment ions resulting from cleavages of carbon-carbon bonds in the fatty acyl moieties were evident at low mass (see Figure 17). Although the carbon number and degree of unsaturation of fatty acyl moieties present in natural mixtures of acyl lipids could be determined the location of the double bonds cannot be established owing to their migration during the CID process.

Tandem mass spectrometry has also been used to characterise the triacylglycerols of human milk following NICI [84 and references therein]. The deprotonated molecular ions of the triacylglycerols were selected using MS1 subjected to CID and then analysis using MS2 to provide information regarding the carbon number and degree of unsaturation of the fatty acyl moieties present. Twenty nine de-protonated triacylglycerol molecular species were studied revealing eleven major acyl moieties including two odd carbon number species. Twenty eight different fatty acids were revealed, including: hexadecanoic acid, which was present in all the spectra, with octadecenoic acid being present in all the unsaturated triacylglycerols. The most abundant triacylglycerol was *sn*-18:1-16:0-18:1 which comprised *ca.*

10% of total triacylglycerol mixture. Although the site of substitution of the fatty acids on the glycerol could be obtained directly from the product ion spectra the locations of the double bonds in the individual fatty acyl moieties could not be deduced.

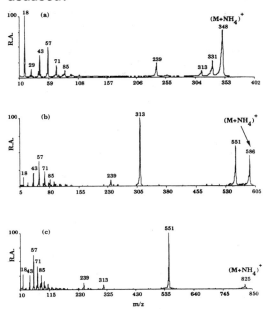

Figure 17. Comparison of the MS/MS product ion spectra of ammoniated pseudo-molecular ions of monopalmitin (a), dipalmitin (b) and tripalmitin (c) using a collision energy of 130eV (reproduced from reference 83 with permission of the American Chemical Society).

Since both the aforementioned studies used a triple quadrupole mass spectrometer only low energy CID could be used. Gross and co-workers have conducted extensive investigations of the high energy CID spectra of fatty acids and acyl lipids ions generated by FAB-MS and analysed using a triple sector mass spectrometer (Figure 18). Free fatty acids dissolved in a matrix of triethanolamine are readily sputtered to yield deprotonated molecular anions, [M-H]⁻. Ions generated in this way have been found to undergo highly specific high energy CID induced 1,4-eliminations of H_2 (Figure 19). These elimination reactions result in losses of an element of C_nH_{2n+2} which starts from the terminus of the alkyl chain and progresses along the entire alkyl chain. Fully saturated fatty acids yield a very smooth envelope of peaks (Figure 20). Gross and co-workers have termed this behaviour 'remote-site fragmentation' [85,86]. A notable advantage of the high energy product ion spectra is that they can be used for the structural investigation of fatty acids. The presence of substituents or unsaturation results in interruptions of the smooth envelope seen in the case of saturated fatty acids (Figure 20). With unsaturated fatty acids the position of double bonds is established by examining high energy CID spectra for an absence of C_nH_{2n+2} losses [85].

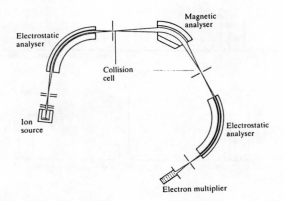

Figure 18. Configuration of the triple sector mass spectrometer used by Gross and co-workers in their studies of the high energy CID FAB-MS/MS behaviour of fatty acids and other acyl lipids.

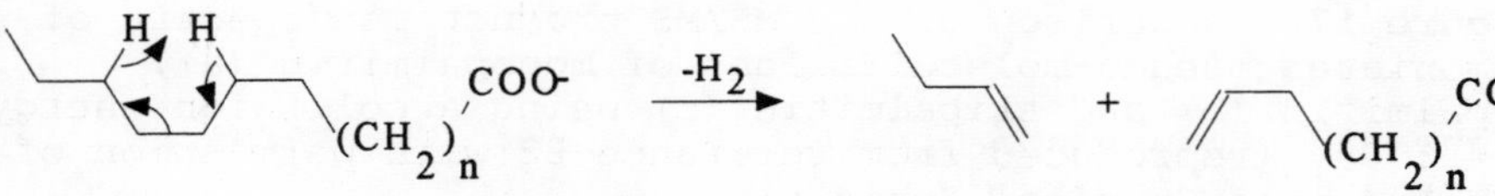

Figure 19. High energy CID induced, 'remote site' fragmentation of an alkyl chain resulting in 1,4-elimination of H_2 [85].

The processes of determining the location of double bonds in polyunsaturated fatty acids is not so straightforward due to preferential expulsion of 45 amu fragments. An alternative strategy involves performing deuterium reduction of the polyunsaturated fatty acid with deutero-diimide in order to label the double bond positions [87]. Their positions are then determined from the mass shifts of product ion peaks in the envelope resulting from the 1,4-elimination of H_2. A further technique that can be used in the case of polyun-saturated fatty acids involves performing high energy CID of cationised species generated at the FAB probe tip [88].

Li^+ was found to yield best results presumably due to the strength of the bond formed in the in the cationised species. The possibility also exist of using high energy CID approach to determine the presence of other structural features of fatty acids, including: alkyl chain branch points, cyclopropane and cyclopropene rings, epoxy and hydroxy substituents [89,90]. The [RCO$_2$]$^-$ ions that appear in the negative ion FAB spectra of phospholipids (see also above) can be selected and subjected to high energy CID to yield structure information concerning their fatty acyl moieties [91]. Compounds that have been studied by this approach include: phosphatidylinositol, phosphatidyl-ethanolamine, cardiolipin, phosphatidic acid and phosphatidylglycerol.

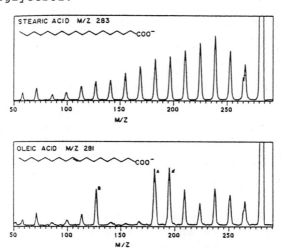

Figure 20. Comparison of the high energy CID FAB/MS/MS spectra of stearic and oleic acids showing the discontinuity in the 'remote site' 1,4-elimination series that can be used to establish the location of the double bonds in unsaturated fatty acids (reproduced with permission of the American Chemical Society from reference 85).

A further refinement of this approach would be to conduct on-line separations in conjunction with tandem mass spectrometry, e.g. GC/MS/MS, SFC/MS/MS or LC/MS/MS, using high energy CID to generate structure information on the fatty acyl moieties present in individual molecular species. Bambagitotti *et al.* [92] have demonstrated that this type of approach is indeed viable through their success in obtaining structure information from unsaturated and branched-chain fatty acid methyl esters in GC/MS analyses involving high energy CID of [RCO$_2$]$^-$ ions generated by NICI. Both CID B/E and CID MIKE spectra were obtained on a double focusing magnetic sector instrument of reversed geometry. When employing a mass spectrometer of EBQQ geometry with low energy collisions the product ion spectra did not contain structurally diagnostic ions resulting from 1,4-elimination H_2.

CONCLUSIONS

Considerable progress has been made in recent years in the field of mass spectrometry with the development of new sampling, ionisation and mass analysis methods. This has greatly enhanced the versatility of the technique for the analysis of all types of organic compound, including those of biological origin. In the lipid field developments in combined chromatographic/mass spectrometric methods, e.g. GC/MS, LC/MS and SFC/MS, offer new possibilities for the analysis of mixtures, particularly of high molecular weight and polar compounds. In many instances the enzymatic or chemical degradations that are commonly employed to release simpler lipid moieties can be avoided since intact acyl lipids can be analysed directly by mass spectrometry. The use of the new combined methods allows previously intractable compositional information to be obtained from highly complex mixtures. Further advances are being brought about through the increased availability of tandem mass spectrometers which provide opportunities for deriving very detailed structure information directly from complex lipids, again without the need for performing prior chemical or enzymatic degradation.

REFERENCES

1.   F.D. Gunstone, in *Lipid Analysis: A Practical Approach* (eds. R.J.Hamilton and S. Hamilton) pp. 1-12, Oxford University Press, Oxford (1992).
2.   R.P. Evershed, in *Lipid Analysis: A Practical Approach* (eds. R.J. Hamilton and S. Hamilton) pp. 113-152, Oxford University Press, Oxford (1992).
3.   R.P. Evershed, in *Lipid Analysis: A Practical Approach* (eds. R.J.Hamilton and S. Hamilton, pp. 263-308, Oxford University Press, Oxford (1992).
4.   R.P. Evershed, In *Gas Chromatography: A Practical Approach* (ed. P. Baugh) pp. 359-391, Oxford University Press, Oxford (1993).
5.   G.R. Waller (ed.) *Biochemical Applications of Mass Spectrometry.* John Wiley, New York (1972).
6.   G.R. Waller and O.C. Dermer (ed.) *Biochemical Applications of Mass Spectrometry*, first supplementary volume. John Wiley, New York (1980).
7.   W.W. Christie, *Gas Chromatography and Lipids: A Practical Guide.* The Oily Press, Ayr (1989).
8.   A. Kuksis and J.J. Myher, in *Mass Spectrometry* (ed. A.M. Lawson) pp. 265-351. Walter de Gruyter, Berlin (1989).
9.   N.J. Jensen and M.L. Gross, *Mass Spectrometry*

Reviews, **6**, 497–536 (1987).

10. D.J. Harvey, *Spectroscopy International Journal*, **8**, 211 (1990).

11. M. Barber, T.A. Merren and W. Kelly, *Tetrahedron Letters*, **18**, 1063 (1964).

12. R.A. Hites, *Methods in Enzymology*, **43**, 348–359 (1975).

13. T. Murata and S. Takahashi, *Anal. Chem.*, **45**, 1816 (1973).

14. T. Murata, S. Takahashi and T. Takeda, *Anal. Chem.*, **47**, 577–580 (1975).

15. T. Murata and S. Takahashi, *Anal. Chem.*, **49**, 728 (1977).

16. A.J. Aasen, H.H. Hofstetter, B.T.R. Iyengar and R.T. Holman, *Lipids*, **6**, 502–507 (1971).

17. T. Murata, *Anal. Chem.*, **49**, 2209 (1977).

18. T. Rezanka, P. Mares, P. Husek and M. Podojil, *J. Chromatogr.*, 355, 265–271(1986).

19. D.E. Games, *Chemistry and Physics of Lipids*, **21**, 389–402 (1978).

20. H.M. Fales, G.W.A. Milne, H.U. Winkler, H.D. Beckey, J.N. Damico and R. Barron, *Anal. Chem.*, **47**, 207 (1975).

21. G.W. Wood, R.C. Charlton and D.G. Hagge, 24th Annual Conference on Mass Spectrometry and Allied Topics, San Diego, p. 385 (1976).

22. N. Evans, D.E. Games, J.L. Harwood and A.H. Jackson, *Biochem. Soc. Trans.*, **2**, 1091 (1974).

23. W.D. Lehmann and M. Kessler, *J. Clin. Chem. Biochem.*, **20**, 893–898 (1983).

24. W.D. Lehmann and M. Kessler, *Biomed. Mass Spectrom.*, **10**, 220–226 (1983).

25. M. Barber, D. Bell, M. Eckersley, M. Morris and L. Tetler, *Rapid Commun. Mass Spectrom.*, **2**, 18–21 (1988).

26. C. Evans, P. Traldi, M. Bambagiotti-Alberti, V. Giannellini, S.A. Coran and F.F. Vincieri, *Biol Mass Spectrom.*, **20**, 351–356 (1991).

27. M.G. Horning, G. Casparrini and E.C. Horning, *J. Chromatogr. Sci.*, **7**, 267 (1969).

28. M. Barber, J.R. Chapman and W.A. Wolstenholme, *Int. J. Mass Spectrom. Ion Phys.*, **1**, 98 (1968).

29. J.J. Myher, A. Kuksis and S.K.F. Yeung, *Anal. Chem.*, **50**, 557 (1978).

30. K. Satouchi, and K. Saito, *Biomed. Mass Spectrom.*, **6**, 396 (1979).

31. R.G. Fenwick. J. Eagles and R. Self, *Biomed. Mass Spectrom.*, **10**, 382–386 (1983).

32. B.N. Pramanik, J.M. Zechamn, P.R. Das and P.L. Bartner, *Biomed. Environ. Mass Spectrom.*, **19**, 164–170 (1990).

33. H. Munster, J. Stein and H. Budzikiewcz, *Biomed. Environ. Mass Spectrom.*, **13**, 423-427 (1986).
34. N.J. Jensen, K.B. Tomer and M.L. Gross, *Lipids*, **22**, 480-(1987).
35. Y.M. Yang, E.A. Sokoloski, H.M. Fales and L.K. Pannell, *Biomed. Environ. Mass Spectrom.*, **13**, 489-492 (1986).
36. K.L. Duffin, J.D. Henion and J.J. Shieh, *Anal. Chem.*, **63**, 1781-1788.
37. J. Rontree, VG Biotech, Altrincham, Cheshire (personal communication of unpublished data).
38. M.C. Prescott, R.P. Evershed and L.J. Goad, unpublished data.
39. W.R. Lusby, M.J. Thompson and J. Kochansky, *Lipids*, **19**, 888-901 (1984).
40. A. Tokumura, Y. Yoshioka, H. Tsukatani and Y. Handa, *Biomed. Environ. Mass Spectrom.*, **13**, 175 (1986).
41. A. Kuksis, *Can. J. Biochem.*, **42**, 407-417 (1964).
42. A. Kuksis, *Can. J. Biochem.*, **42**, 419-430 (1964).
43. A. Kuksis, L. Marai and D.A. Gornall, *J. Lipid Res.*, **8**, 352-358 (1967).
44. A. Kuksis, O. Stachnyk and B.J. Holub, *J. Lipid Res.*, **10**, 660-667 (1969)
45. P. Mares, *Prog. Lipid Res.*, **27**, 107-133 (1988).
46. P.P. Schmid, M.D. Muller and W. Simon, *JHRC & CC*, **2**, 675-676 (1979).
47. R.P. Evershed, M.C. Prescott and L.J. Goad, *Rapid Commun. Mass Spectrom.*, **4**, 345 (1990).
48. R.P. Evershed and L.J. Goad, *Biomed. Environ. Mass Spectrom.*, **14**, 131 (1987).
49. R.P. Evershed, M.C. Prescott, L.J. Goad and H.H. Rees, *Biochem. Soc. Trans.*, **15**, 175 (1987).
50. R.P. Evershed, V.L. Male and L.J. Goad, *J. Chromatogr.*, **400**, 187 (1987).
51. R.P. Evershed and L.J. Goad, *Biomed. Environ. Mass Spectrom.*, **16**, 169 (1989).
52. R.P. Evershed, M.C. Prescott, N. Spooner and L.J. Goad, *Steroids*, **53**, 285 (1989).
53. R.P. Evershed and L.J. Goad, in Atta-ur-Rahmann (ed.) *Studies in Natural Product Chemistry*, Vol. 9, pp. 447-486, Elsevier, Amsterdam (1991).
54. R.P. Evershed, M.C. Prescott and L.J. Goad, *J. Chromatogr.*, **590**, 305 (1992).
55. S.G. Wakeham and N.M. Frew, *Lipids*, **17**, 831-843.
56. T.A. Roy, F.H. Field, Y.Y. Lin and L.L. Smith, *Anal. Chem.*, **51**, 272-278.
57. L. Dyas, M.C. Prescott, R.P. Evershed and L.J. Goad, *Lipids*, **26**, 536 (1991).
58. T. Rezanka, *J. Chromatogr.*, **598**, 291-226 (1992).
59. S.R. Lipisky and M.L. Duffy, *J. HRC & CC*, **9**, 725

(1986).

60. A. Kuksis, J.J. Myher and P. Sandra, *J. Chromatogr.*, **500**, 427 (1988).

61. J.J. Myher, A. Kuksis, L. Marai and P. Sandra, *J. Chromatogr.*, **452**, 93 (1988).

62. T. Ohshima, H.S. Yon and C. Koizumi, *Lipids*, **24**, 535 (1989).

63. R.P. Evershed and M.C. Prescott, *Rapid Communications in Mass Spectrometry*, submitted for publication.

64. R.W.P. Scott, C.G. Scott, M. Munroe and J. Hess, Jr., *J. Chromatogr.*, **99**, 395 (1974).

65. O.S. Privett and W.L. Erdahl, *Chem. Phys. Lipids*, **21**, 361-387 (1978).

66. W.H. McFadden, H.L. Schwartz and S. Evans, *J. Chromatogr.*, **122**, 389 (1976).

67. W.H. McFadden, D.C. Bradford, D.E. Games and J.L. Gower, *Amer. Lab.*, October (1977) 55-64.

68. N.W. Rawle, R.G. Willis and J.D. Baty, *Analyst*, **115**, 521-523 (1990).

69. A. Kuksis, L. Marai, J.J. Myher and D. Pino, in *Chromatography of Lipids in Biomedical Research and Clinical Diagnosis*, ed. A.Kuksis, pp. 403-440, Elsevier, Amsterdam (1987).

70. R.P. Evershed, in *Handbook of Derivatives for Chromatography* (K. Blau and J.M. Halket, eds), pp. 52-108, John Wiley and Sons, Chichester (1993).

71. M.L. Blank, M. Robinson, V. Fitzgerald and F. Snyder, *J. Chromatogr.*, **298**, 473-482.

72. L. Marai, A. Kuksis, J.J. Myher and Y. Itabashi, *Biological Mass Spectrom.*, **21**, 541-547 (1992).

73. A. Kuksis, L. Marai and J.J. Myher, *Lipids*, **26**, 240-246 (1991).

74. H.-Y. Kim and N. Salem, *Anal. Chem.*, **58**, 9-14 (1986).

75. H.-Y. Kim and N. Salem, *Anal. Chem.*, **59**, 722-726 (1987).

76. A.I. Mallet and K. Rollins, *Biomed. Environ. Mass Spectrom.*, **13**, 541-543 (1986).

77. R.T. Dobrowsky, R.D. Voyksner and N.C. Olson, *Am. J. Physiol.*, **260**, H1455-H1465 (1991).

78. T. Kasama, Y. Seyama, *Nippon Iyo Masu Supekutoru Gakkai Koenshu*, **15**, 141-144 (1990).

79. H. Kallio, P. Laakso, R. Huopalahti, R.R. Linko and P. Oksman, *Anal. Chem.*, **61**, 698-700 (1989).

80. M.V. Merritt, D.M. Sheely and V.H. Reinhold, *Anal. Biochem.*, **193**, 24034 (1991).

81. K. Matsuura, M. Takeuchi, K. Nojima, T. Kobayashi and T. Saito, *Rapid. Commun. Mass Spectrom.*, **4**, 381-383 (1991).

82. E. Huang, J.D. Henion and T.R. Covey, *J. Chromatogr.*, **511**, 257-270 (1990).

83. K.L. Duffin, J.D. Henion and J.J. Shieh, *Anal. Chem.*,

    63, 1781-1788 (1991).
84. G.J. Currie and H. Kallio, *Lipids* **28**, 217 (1993).
85. K.B. Tomer, F.W. Crow and M.L. Gross, *J. Amer. Chem. Soc.*, **105**, 5487-5488 (1983).
86. N.J. Jensen, K.B. Tomer and M.L. Gross, *J. Amer. Chem. Soc.*, **107**, 1863-1868.
87. N.J. Jensen, K.B. Tomer and M.L. Gross, *Anal. Chem.*, **57**, 2018-2021 (1985).
88. J. Adams and M.L. Gross, *Anal. Chem.*, **59**, 1576-1582 (1987).
89. N.J. Jensen and M.L. Gross, *Lipids*, **21**, 362-365 (1986).
90. K.B. Tomer, N.J. Jensen and M.L. Gross, *Anal. Chem.*, **58**, 2429-2433 (1986).
91. N.J. Jensen, K.B. Tomer and M.L. Gross, *Lipids*, **22**, 480-489.
92. M.Bambagiotti A., S.A. Coran, F.F. Vincieri, T. Petrucciani and P. Traldi, *Organic Mass Spectrometry*, **21**, 485-488 (1986).

Analysis of Lipid Structure by Neutron Diffraction

M. J. W. Povey

PROCTER DEPARTMENT OF FOOD SCIENCE, THE UNIVERSITY,
LEEDS LS2 9JT, UK

1.   INTRODUCTION

There are a great number of similarities between Neutron Diffraction and X-ray diffraction, however it is not possible to carry out Neutron Diffraction experiments on the bench top with equipment costing a few tens of thousands of pounds. The two main neutron diffraction facilities in Europe, at the Institut Laue Langevin (ILL) in Grenoble, France and the ISIS at the Rutherford Appleton Laboratories in the U.K. cost many millions of pounds both to build and to run. So neutron diffraction will be used for measurements impossible to make in any other way.

Whilst lipids have been widely studied using neutrons, attention has concentrated on biological membranes rather than lipids themselves. Pioneering work on the properties of membranes and membrane components employed small angle diffraction from lipid bilayers and biological membranes, small angle scattering from aqueous dispersions of membranes and membrane components and Quasi-elastic and inelastic scattering measurements of the dynamical properties of membrane components.[1]

Studies of phospholipids began in the early 1970s, facilitated by the provision of neutron diffraction facilities in the European centres at Harwell and the Institut Laue-Langevin in Grenoble. Materials such as Synthetic and Egg Lecithin, Cholesterol, and Myelin received considerable attention, studied for their role in the formation of membranes. This interest has continued unabated to this day[2].

This author is unaware of any previous neutron studies of the triacylglycerols. Many problems of triacylglycerol structure are susceptible to neutron study, Larssons paracrystal hypothesis is an example of one which has failed to yield to other methods.[3] The time has come to add neutrons to the armoury of techniques available for the study of the simple lipids. In this chapter, neutron scattering for studying lipids will be exemplified by experiments carried out by the author and his collaborators.

2.   EXPERIMENTAL EQUIPMENT

Most of the experimental work to be described was carried out at the ILL using the D17 and D11 diffractometers. The basic elements of the diffractometer are shown in Figure 1, the layout is similar to that of an X-ray diffractometer. The neutron source is a nuclear reactor.

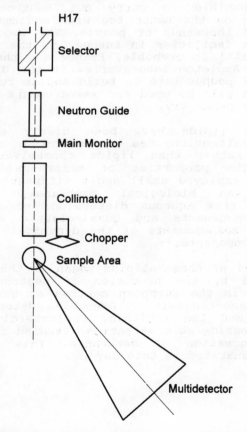

H17

Selector

Neutron Guide

Main Monitor

Collimator

Chopper

Sample Area

Multidetector

Figure 1.Diagram of the D11 Neutron Diffractometer at the ILL, Grenoble.

The neutrons are 'thermalised', i.e. slowed down to reduce their energy and hence increase wavelength before being guided to the diffractometer. Wavelength selection is carried out using a rotating spiral groove which selects neutrons of the desired velocity and hence wavelength. Wavelengths between 0.2 Å and 40 Å are available.

The sample is placed in the neutron path and the neutron intensity measured as a function of angle, using the rotating multidetector. Neutron intensity is a function of the number of neutrons arriving at any point in the detector. Full details of the ILL equipment are given in Maier.[4]

3. THE PRINCIPLES OF NEUTRON DIFFRACTION

A good general introduction to neutron diffraction for the analysis of biological materials is given by B. Jacrot.[5] The basics of neutron diffraction will be familiar to anyone who has worked with X-ray diffraction. For the sake of explanation, let us assume that we have a single particle which is rigidly fixed (Figure 2). A plane parallel wave front moves from left to right, representing a single neutron. The amplitude of the neutron, i.e. the square root of the probability that it will trigger a detector at the point z is

$$\Psi_i(z) = exp(ik_o z) \tag{1}$$

where

$$k_o = \frac{2\pi}{\lambda}. \tag{2}$$

Here $\Psi_i(z)$ is the probability amplitude of encountering an incident neutron at a distance z from the neutron source, k_o is the amplitude of the wave vector of the neutron and is inversely proportional to the wavelength λ.

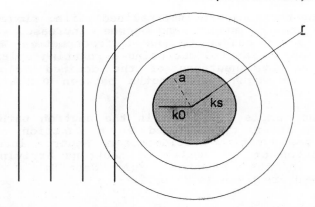

Figure 2 . Incoming plane wave scattering from a spherical object. The parallel lines represent lines of constant phase for the incoming wave and the circles represent circles of constant phase for the scattered wave. The result is summed at the point r.

Experimentally, it is found that the scattering will be predominantly spherically symmetric with the scattered intensity taking the form

$$\Psi_s(r) = -\frac{b}{r} \, exp(i\overline{k}_s \bullet \overline{r}_s) \tag{3}$$

where $\overline{k}_s$ is the wave vector of the scattered neutron, since we assume elastic scattering the amplitude of $\overline{k}_s$ will equal k_0. $\overline{r}_s$ is the direction in which the neutron is scattered and r is its amplitude, i.e., the distance of the detector from the sample.

The scattered intensity decays according to an inverse square law. The resultant probability amplitude of a neutron appearing in the detector at the point r is just the sum of the incident and scattered probabilities:-

$$\Psi_{result} = \Psi_i + \Psi_s. \tag{4}$$

Coherent scattering patterns arise from the spatial correlations between coherent scattering sites. If energy transfer is neglected then the scattering is said to be elastic. In this case the coherent scattering cross section per atom is

$$\frac{d\sigma}{d\Omega}(\underline{Q}) = \frac{1}{N} \left| \Sigma_R \, b_R \, exp(i\underline{Q}.\underline{R}) \right|^2 \tag{5}$$

where N is the number of scattering nuclei, b_R is the coherent scattering length of a nucleus at position vector R, and Q is the scattering (momentum transfer) vector defined by the difference between incident and scattered wave vectors with

$$|\underline{Q}| = \frac{4\pi \sin \theta}{\lambda} \tag{6}$$

where 2θ is the scattering angle and λ is the neutron wavelength. $\frac{d\sigma}{d\Omega}(\underline{Q})$ is related directly to the observable intensity distribution of scattered neutrons, $I(Q)$ by experimental factors which are beyond the scope of this review.

The real space dimension, d, observed at a momentum transfer of Q is

$$d = 2\pi / Q \tag{7}$$

4. A COMPARISON OF NEUTRON AND X-RAY DIFFRACTION

The X-ray and neutron coherent scattering amplitudes of some atoms are given in Table 1.

There is no regularity of the neutron amplitudes as there is with X-rays. This is because neutron scattering is the result of nuclear forces and is dominated by a process known as resonant scattering in which the incident neutron and the target nucleus combine to momentarily form a compound nucleus. The scattering amplitude is then determined by the energy levels of the compound nucleus and consequently there is little regularity with the periodic table. Another consequence of resonant scattering is that the scattering amplitude can be negative as for hydrogen. There is nothing analogous for X-rays.[1] Because neutrons are scattered by nuclei of atoms, the scattering amplitudes are different for different isotopes of the same element. Isotopic substitutions therefore provide a direct means of isomorphous changes of scattering amplitudes. The large difference between the coherent scattering amplitudes of hydrogen and deuterium (Table 1) is particularly useful in the study of biological materials. Utilisation of this large difference is a basic motivation for neutron studies of biological materials.

Incoherent Scattering Cross-section.

Neutrons exhibit an incoherent scattering cross-section, due to nuclear spin. which is absent in ^{12}C or ^{16}O but is very important for ^{1}H. It generates an isotropic background. This background is subtracted

Isotope	Molecular Weight	X-rays[a] $f_x \cdot 10^{12}$ cm	Neutrons[b] $b_{coh} \cdot 10^{12}$ cm
^{1}H	1.008	0.28	-0.374
^{2}H (D)	2.104	0.28	+0.667
^{12}C	12.000	1.69	+0.665
^{13}C	13.003		+0.6
$(^{12}$C+^{13}C)$^+$	12.01		+0.664
^{14}N*	14.003		+0.94
^{16}O*	15.999	2.25	+0.580
^{23}Na*	22.990		+0.36
^{27}Al	26.982	3.65	+0.350
^{28}Si	28.086	3.95	+0.420

* Natural abundance greater than 99.5%
+ Natural abundance of C^{12} is 98.89% and for C^{13} it is 1.11%.
a The form factor, f_x, for X-rays is given at $\sin\Theta = 0$.
b The coherent scattering length per atom at a neutron wavelength of 1.08 Å. The coherent cross section is given by $4\pi b^2$.

Table I Values of coherent neutron scattering lengths for some isotopes. [67]

from the sample signal by using a calibration material such as water.

A second type of incoherence arises from elements which consist of more than one isotope randomly distributed in the sample at the atomic positions of the element.

Biological materials generally have much better stability in neutron beams than in X-ray beams because of their small cross-section. This is a major advantage of the neutron technique. Neutrons penetrate deeply into samples before suffering significant attenuation. As a result, relatively large samples can be used in neutron diffraction while still retaining sensitivity to all effects throughout the material under study and not just the surface regions. It also eliminates possible sample size effects when studying order.

The Coherent Scattering Cross-section

The neutron scattering length. This does not depend on scattering angle. Nuclei may be considered to be points. The coherent scattering cross-section for neutrons is equivalent to the X-ray form factor but is independent of angle. In the case of X-rays, the form factor varies with angle and must be calculated for each element.

Neutron scattering length varies little with atomic number. Unlike X-rays, whose scattering length is much smaller for H than for C, O or N, the scattering of neutrons is almost independent of atomic number. However, the phase of the scattering is reversed in the case of the proton, resulting in a large contrast between protons and other nuclei. Thus protons contribute much more to neutron scattering than to X-ray scattering.

In the case of X-rays, the interaction with matter is *via* the electronic structure of the individual atoms. For practical purposes this means that in materials such as lipids the carbon and oxygen atoms dominate the X-ray diffraction pattern. Little or no information can be obtained on hydrogen positions, because the so-called "Form factor" is so small.

Because X-rays are easily absorbed by matter, only small samples can be used before multiple scattering and self shielding effects begin to deteriorate the quality of information in the scattered beam.

Contrast Matching There is a $180°$ phase shift between the Hydrogen and Deuterium scattering lengths. Therefore the replacement of a proton by a deuteron (Neutron + proton) is a major modification to the scattering. This makes contrast matching possible, an important advantage for neutron scattering.

It is possible to study the distribution of protons in the system this way. The proton/deuteron ratio can be altered be substituting deuterium oxide for water in the synthesis of the materials under study, thus allowing different parts of a molecule to be studied.

Wavelength. Neutrons from a reactor source can be selected with wavelengths of any value and with a narrow distribution. Neutrons can be produced at longer wavelengths (up to 40Å) than X-rays and therefore larger structures can be examined.

A wavelength of 1.54 Å (Cu K_α) is the most commonly available in laboratory X-ray equipment. A short wavelength is good for observing the high-angle ($15° < 2\theta < 30°$) reflections from the short spacings, but somewhat lacking when probing the long spacings of 30 Å and above. Long spacings produce reflections at low momentum transfer (angles of $2\theta < 5°$) and separation of the diffraction peak from the main beam is difficult.

Definition of angular position and shape of peaks is often unreliable. Observation of both angular regions must be performed with different angular resolutions, which renders intensity comparisons between long and short spacing effects subject to reduced reliability. For example, it is difficult to examine both the long and short spacing regimes in triacylglycerols in one experiment using X-rays.

On the D16 instrument at Grenoble, d-spacings between <1Å and 125Å can be recorded in the same experiment.[8] The longer wavelengths available with neutrons shift observations of any particular feature to higher angles. By using neutrons with $\lambda = 1.54$ Å, low angle reflections are shifted from about 3°, for example, to 9°. That puts the feature well out of interference from the main beam and allows for the use of a single resolution over the angular region of interest.

5.   THE DISTINCTION BETWEEN EQUIPMENT AT THE TWO EUROPEAN CENTRES

The major differences between the two main European neutron sources are summarised in Table 2.

Another major difference between the two sources at the time of writing is that the ILL source is out of commission and is not due back in operation until some time in 1994.

ILL	ISIS
Continuous beam	Pulsed beam
$0.003 < Q < 1$ Å^{-1} (H17)	$0.005 < Q < 50$ Å^{-1}
$0.2 < \lambda < 40$ Å	$0.5 < \lambda < 6.5$ Å
Reactor	Proton accelerator and spallation source

Table 2 A comparison of the ILL and ISIS Neutron sources

6.   APPLICATIONS

Neutron diffraction is well suited to the study of biological materials.[1] Small angle scattering studies have provided important information about structure in lipid crystals, bi-layers and membranes.[5,6,1] In this review recent studies on lipid systems are described[7,8] which illustrate well the possibilities for studying lipid systems such as triglycerides in emulsions.

Neutron diffraction studies of liquid and crystalline trilaurin.

Our interest lay in the possibility of order in the liquid state of triglycerides. This has been of considerable interest in the past and experimental evidence for it has been offered by several authors.[3,9-12] We turned to neutron scattering to elucidate the expected order.

Measurement of the width of X-ray diffraction reflections led Larsson[3] to propose that lamella units increase in size continuously as temperature is decreased towards the melting point, until crystallisation finally occurs. Larsson determined the persistence length in the lamellae to be about 200Å. Other data exist that support the hypothesis of phase changes in liquid crystalline order in triglyceride melts. However, none of the structural changes suggested in the liquid state have been reported as being recorded by calorimetric techniques so if they exist they must be weakly energetic.

Materials and Methods. Two forms of trilaurin were synthesised. The first was made from glycerol-H_8 and deuterated lauric acid-D_{23} [$CD_3(CD_2)_{10}COOH$]. A compound is produced with only 5 protons per molecule located at the glycerol part. The fatty acid chains were therefore fully deuterated and contain 69 deuterium atoms. This form of trilaurin will be referred to as LLL-HD.

The second form of trilaurin was made from fully deuterated glycerol-D_8 and a lauric acid, which was deuterated in the hydrogen positions adjacent to the COOH group of the acid [dodecanoic-2, 2-D_2 acid, $CH_3(CH_2)_9$-CD_2COOH]. The resulting molecule is one in which all hydrogen positions in the vicinity of the glycerol were deuterated. The fatty acid chains were protonated beyond the first acyl carbon. This second form of trilaurin will be referred to as LLL-DH.

The apparatus used for our experiments was the D-16 diffractometer at the Institut Laue-Langevin (ILL), Grenoble, France. D-16 is a two circle diffractometer which is depicted in Figure 1. It provides a neutron wavelength of $\lambda = 4.5$Å from a pyrolytic graphite monchromator system. The detector consists of a two-dimensional wire grid system giving a 64x16 matrix. The detector can be placed at various distances from the sample, around which it rotates.

The angular range accessible lies between about $3 < 2\theta[°] < 120$. when placed at 1 m from the sample, this gives a momentum transfer range of about $0.05 < Q$ [Å] < 2. Therefore d spacings as large as 125Å can be recorded in the same experiment as those less than 1Å. Triglycerides display characteristic sets of d-spacings between these two extremes and therefore the technique is well adapted to their study.

The same experiments were performed on each of the two samples. Samples were melted for a few minutes at slightly above 60°C. The temperature was set at 60°C and a diffraction pattern recorded over a period of about an hour. The temperature was then reduced to ambient (about 25°C) over several hours to form a solid phase by slow crystallisation. A diffraction pattern was then recorded from the solid. The samples were remelted and then cooled in stages from 60°C.

Patterns were recorded at several temperatures, at about 5°C intervals, between 60°C and about 35°C, by allowing an hour or so for the diffraction pattern to be scanned at each temperature. The melting point for the solid in the β phase was ≈45°C so that when the liquid cools below 45°C it supercools.

The cooling process between each temperature was also slow. Crystallisation eventually occurred at 32°C. Diffraction patterns were recorded several times at each temperature. After crystallisation the temperature was allowed to fall to ambient. The samples were then slowly reheated to around 70°C over a period of 18hours. Diffraction patterns were recorded every 15 minutes or so throughout this process.

The resultant diffraction patterns for LLL-HD in both solid and liquid states are shown in Figure 3.

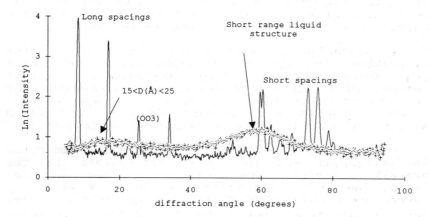

Figure 3. Neutron diffraction patterns obtained from LLL-HD. In the solid state after slow crystallisation at room temperature from the melt (full line, sharp reflections), and the molten state at 61°C (broken line, broad feature at about 2θ of 60°). All diffraction patterns obtained from LLL-HD in any of the liquid states were identical (from Cebula et al [7]).

For the low temperature case (ambient temperature about 25°C) the diffraction obtained is strong in both low and high angle regions. For LLL-HD, the neutron diffraction pattern resembles the conventional high-angle pattern seen with X-rays.

The neutron diffraction pattern has more diffraction lines because, with deuterium, there are more scatterers in the unit cell. In the crystal it is known that a layer structure is formed and a series of (001) basal reflections, seen in the data, are not unexpected. Clear reflections appear from (001) to (004). These reflections correspond to a layer spacing of about 33Å which corresponds to the double layer structure expected for trilaurin. The high-angle reflections are also intense and show a peak at 4.55Å (2θ about 60°), characteristic of the β polymorphic form.[13]

For the β phase in triglyceride crystals, the glycerol parts of neighbouring molecules reside in a single plane, and fatty acid chains project away from that plane and are approximately parallel to each other. For LLL-HD molecules, all the hydrogen atoms of the fatty acid chains are deuterated. Both low and high angle sets of reflections have comparable intensity.

As deuterium has such a high cross-section, the diffraction from chain-chain organisation, or short-range order, makes a large contribution. Little contrast in scattering is evident in proceeding from one layer to the next parallel to the long spacing. The LLL-DH sample on the contrary shows a diffraction pattern dominated by the layer organisation (Figure 4).

The more heavily deuterated glycerol now dominates the scattering and the (001) reflections in the crystalline material are much more intense than for LLL-HD. The high angle reflections are now virtually absent, even when plotted on the logarithmic scale. When the samples are melted (45°C), all the sharp peaks disappear, to be replaced in the LLL-HD case with a broad and much less intense feature in the high angle region. This is characteristic of pair correlations of atoms in an isotropic liquid with average d-spacings larger than in the solid. Identical patterns to those in the melt were obtained in the supercooled liquid.

There was absolutely no evidence of any crystalline type order in the supercooled melt (below about 45°C) at 39°C, down to about 33°C just above the crystallisation temperature. Nor was there any evidence of crystalline type order just above the melting point of the solid (45°C). We conclude from these studies that the classical picture of structure proposed for liquid triglycerides cannot be sustained by the neutron diffraction evidence.

If the lamellae postulated by Larsson exist with a well defined structure, they should produce reflections for both long and short spacings. The selective deuteration of the molecule would amplify any such features. We can find no evidence for lamellae, and any such ordered material must occupy less than 1% of the volume of liquid present. The interaction time between a neutron and the sample is far shorter than the molecular diffusion times in the melt so order will be detected, even if it is transient.

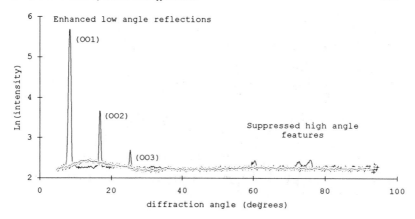

<u>**Figure 4 Neutron diffraction patterns obtained from LLL-DH.**</u> In the solid state after slow crystallisation at room temperature from the melt (full line, sharp reflections), and the molten state at 61°C (broken line, broad feature at about 2θ of 60°). All diffraction patterns obtained from LLL-DH in any of the liquid states were identical.

In Figure 5 a structure for the liquid state in trilaurin is postulated[7] which is consistent with the neutron data. It is a nematic-like structure in which layers are absent. The analogy of pencils in a box is not inappropriate here.

The sketch below gives an idea of the spatial arrangement of molecules with respect to one another, it de-emphasises the internal structure of the trilaurin molecule. Although the fatty acid moieties are depicted as straight lines, this is not to suggest that they are in fact rigid. Since this is a liquid this structure represents a frozen moment from a continually changing structure.

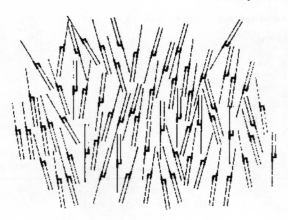

Figure 5 Schematic diagram of postulated structure for the liquid state in trilaurin [7]

Small angle neutron scattering (SANS) from voids in crystalline trilaurin[8]

In an earlier experiment to find evidence of order in trilaurin we used the D17 diffractometer at the ILL to investigate evidence of scattering from the paracrystals postulated by Larsson[3]. Larsson had suggested that, just above the melting point, the persistence length in the lamellae should be 20 nm. D17 differs from D11 in that it specialises in measuring small angle scattering, hence allowing the investigation of structures of up to several thousand Å.[4] We used this instrument in an attempt to obtain the size of the hypothesised lamellae. The results are summarised in Figure 6.

The material used in this experiment was synthesised from glycerol-H (i.e. ordinary protonated glycerol) and deuterated lauric acid. This deuterated sample has an increased scattering cross section with respect to neutrons, thus improving the chances of detecting order. All phase transitions of the trilaurin were located prior to the Small Angle Neutron Scattering (SANS) experiment using Differential Scanning Calorimetry (DSC).

The deuterated trilaurin was heated to 65°C, well above its β melting point (43°C), for about 30 min or so. This was to ensure that any crystalline order present was destroyed. It was then thermostatted at 34°C (i.e. above the α crystallisation point, 15°C) and SANS measurements were carried out every 20 minutes until the trilaurin crystallised (≈ 4 hr).

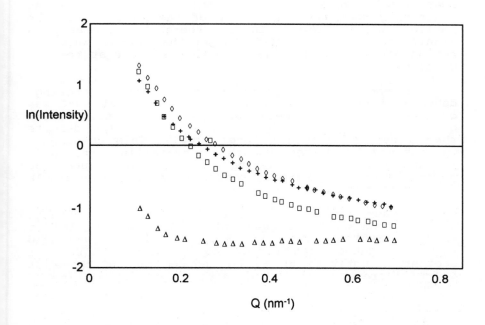

Figure 6. Variation of relative intensity with momentum transfer for trilaurin melt (Δ), trilaurin crystallised at 34°C (□), quenched trilaurin (+) and annealed trilaurin (◊), $Q = 4\pi \sin \theta / \lambda$.

The melt showed no discernible scattering above that from an isotropic liquid i.e. there was no angular dependence in the scattering pattern (Figure 6). The slight increase in scattering below Q values of about 0.2 nm^{-1} is caused by some of the incident beam falling on the detector at small angles. There was also no evidence of a change in the scattering pattern from the melt over the 4 hr period.

These results suggest that there is either no crystalline order in the super cooled melt or that the SANS used in this experiment was not sensitive enough to detect it. The latter would be the case if the density contrast between the ordered and disordered regions in the melt was less than 1%.

Once the trilaurin had completely crystallised (at 34°C), strong incoherent scattering appeared (Figure 6). The intensity of the scattered signal rose to over 100 times that in the melt. This was attributed[8] to the presence of voids (or defects) in the crystalline structure. This is because voids have a large contrast in magnetic moment with deuterated trilaurin. Figure 6 shows scattering with correlation lengths $(2\pi/Q)$ between 60 nm (at high Q) and 400 nm (at low Q). The scattering amplitude is greatest at low Q values indicating that there are more of the larger than the smaller sized voids. Indeed there is probably a significant number of voids larger than 400 nm which do not appear in Figure 6 because they scatter at too low Q values.

To confirm our hypothesis of defect scattering, deuterated trilaurin melt held for an hour at 50°C was quenched in ice. In metal and polymer crystallisation quenching produces a higher number of smaller defects than a sample allowed to cool slowly. This phenomenon was observed in the quenched crystalline trilaurin (Figure 6). The scattering at large Q values was greater from the quenched sample than from the sample allowed to crystallise at 34°C.

Further confirmation that the voids were defects was obtained when the quenched sample was annealed by raising the temperature to 35°C, at which it was held for 30 minutes then cooled to 25°C. The number of scattering entities at 60 nm reduced whilst those at 400 nm increased, consistent with disproportionation of defects.

Voids of this size range are not altogether surprising if they are regarded as defects in the formation of crystals whose unit cell size in the case of trilaurin is about $0.35 \times 3 \times 0.2$ nm^3. Although the SANS measurements carried out in this work only give a qualitative indication of the size distribution of the voids, they nevertheless, clearly demonstrate the usefulness of SANS for characterising voids in crystalline triglycerides.

There have been no other attempts, so far as the author is aware, to characterise voids in crystalline fats. However, the presence of voids would explain the high attenuation of ultrasound in solid triglycerides[14] and the anomalous expansion of simple even triglycerides observed by Hvolby[15]. Defects in crystalline triglycerides, caused by the incorporation of surfactants in the melt, are believed to play an important role in regulating polymorphic transitions[16].

Other applications

Neutron Reflectivity. The CRISP instrument at the Rutherford-Appleton Laboratory near Didcot, England, is a dedicated reflectometer which uses a fixed angle of incidence and a pulsed, polychromatic, neutron beam. Specularly reflected neutrons are received by a single detector and their wavelengths are analysed by time of flight. Perpendicular wave-vector transfer is scanned by measuring reflectivity as a function of wavelength.[17]

A polymer, adsorbed at an interface can be studied by changing the ratio of proton to deuteron in the oil and aqueous phases so that the neutron contrast in the two bulk phases is matched. In this case reflection is largely confined to the interface, and accordingly information about the volume fraction profile of protein adsorbed at the oil information can be obtained.

For the CRISP instrument it is necessary to deploy the interface as a macroscopic film and Dickinson co workers[17] describe an experimental apparatus for doing this. However, if D11 at the ILL is used it is possible to employ the same concept of contrast matching to obtain information about polymer adsorbed at the interface in an emulsion, a technique which this author and his collaborators will elaborate on at a later date.

7.   REFERENCES

1   D.L. Worcester, in 'Biological Membranes, Vol 3'(D. Chapman and D.F.H. Wallach, eds.) Academic Press, London, 1976, Chapter 1, p1.
2   Institut Laue Langevin 'Annual Report 1992' ILL, Grenoble, 1992, p78
3   K. Larsson, *Fette. Seifen Anstrichm.*, 1972, 74, 136.
4   B. Maier, 'Neutron Research Facilities at the ILL High Flux Reactor', Institut Laue Langevin, Grenoble,1983.
5   B. Jacrot *Rep. Prog. Phys.*, 1976 39 911.
6   G.E. Bacon, ' Neutron Diffraction', 3rd Edition, Clarendon Press, Oxford, 1975 p38.
7   D. J. Cebula, D. J. McClements, M.J.W. Povey and P.R.Smith, *JAOCS*, 1992, 69 130.
8   D.J. Cebula, M.J.W. Povey and D.J. McClements, *J.Am.Oil Chemists' Soc.*, 1990, 67, 76.
9   L.W. Phillips, *Trans. Farad. Soc.*, 1964, 60, 1873.
10   P.T. Callaghan and K.W. Jolley, *J. Chem. Phys.*, 1977, 67, 4773.
11   I.T. Norton, C.D. Lee-Tuffnell, S. Ablett and S.M. Bociek, *J.Am.Oil Chemists' Soc.*,. 1990, 67, 76.

12   C.V. Oerr and H.J. Yarwood, <u>J. Phys. Chem.</u>, 1956,
      <u>60</u>,1265.
13   D. Chapman, 'The Structure of Lipids', Methuen &
      Co. Ltd, London, 1965, 266.
14   D.J. McClements., 'The use of ultrasonics for
      characterising fats and emulsions.', Ph. D.
      thesis, University of Leeds, Leeds, 1988.
15   A. Hvolby, <u>J. Am.Oil Chemists' Soc.</u>, 1974, <u>51</u>, 50
16   J.S. Aronhime, S. Sarig and N. Garti, <u>J.Am.Oil
      Chemists' Soc.</u>, 1988, <u>65</u>, 144
17   E. Dickinson, D.S. Horne, J.S. Phipps and R.M.
      Richardson, <u>Langmuir</u>, 1993, <u>9</u>, 242.

Purity Criteria in Edible Oils and Fats

J. B. Rossell

LEATHERHEAD FOOD RA, RANDALLS ROAD, LEATHERHEAD,
SURREY KT22 7RY, UK

1. INTRODUCTION

A review is presented of edible vegetable oil purity criteria developed at the Leatherhead Food Research Association. Most of the work involved accurate determination, by modern GLC techniques, of the fatty acid compositions of oils extracted in the laboratory from oil- seeds of known origin and history. All of the main production areas throughout the world were represented in the collection of over 600 samples of commercial oilseeds No botanical curiosities or hand-picked specimens were included as the work related to commercially available edible oils. The fatty acid compositions of the major vegetable oils are reviewed, and the influence this had on the revision of Codex Alimentarius Fats and Oils Specifications is discussed. The development of purity criteria based on the composition of fatty acids at the triglyceride 2-position, triglyceride compositions by high-temperature GLC, sterol compositions and tocopherol concentrations is also reviewed.

In the case of maize oil a significant new development is the authentication of the oil by stable carbon isotope ratio measurement. The possibilities of this exciting new technique are reviewed in the light of 42 results on maize oils of various origins, together with over 60 results on a selection of other oils and fats.

2. BACKGROUND

When the author joined the Leatherhead Food RA in 1980, there were a number of edible oil purity problems in the world. Some of these, such as the contamination of beef tallow with lard, which is a cause of great concern to Muslims and Jews still exist, at least to the extent that we cannot yet prove absence of lard with sufficient rigor to satisfy their religious principles. However, several problems have been solved to the satisfaction of the international oils and fats trade, and these will be discussed paying particular attention to the major vegetable oils.

These oil purity problems included: blending of palm stearin and palm olein with each other or with palm oil in Singapore to give the so-called "Singapore Cocktail", which was of variable quality; the suspected presence, in some groundnut oils from South America, of small amounts (of, say, 5%) of other liquid oils such as soya-bean oil; the blending of rape and soya oils in Europe, rape being usually cheaper but difficult to detect at levels of less than 10% in soya; the sale of corn oil imported into the United Kingdom from Continental Europe at impossibly low prices, giving grave concern about its purity; and the co-mingling of cottonseed oil imported into Egypt with palm olein produced in Malaysia, causing serious problems.

This last problem was a case of major fraud and deserves greater attention. One story told is that a Singapore dealer negotiated a contract to sell a large consignment of, perhaps, 10,000 tons of cottonseed oil to the Egyptian Government. For the first shipment of ca 5,000 tons, he had about 4,250 tons of Australian cottonseed oil, but found difficulty securing the balance. He therefore 'diluted' or 'extended' the 4,250 tons of cottonseed oil with about 750 tons of palm olein, the liquid fraction from palm oil. This was delivered to Egypt apparently without any problem. As a result, when the time came for him to deliver the balance of the contract he used a much larger palm olein content; some say he might have reversed the ratio of cottonseed oil to palm olein. Fortunately, this fraud was detected and the fraudster went to prison. If he had been successful it would have gained him an additional US$14M. The story was widely reported in the Malaysian newspapers, as the Malaysian Government, which had provided the original palm olein, but had been blameless with regard to the fraud, quite naturally wished to distance itself from the incident.

Cottonseed oil has been studied in detail, and some of this work has been reported. [1,2].

The Authenticity Project

The authenticity project was established at the Leatherhead Food RA with the objective of establishing new and up-to-date purity criteria for the world's major vegetable oils to ensure consumer protection and promotion of fair trade by prevention of fraud.[3,4] The project was financed by the Federation of Oils Seeds and Fats Association Ltd. (FOSFA International); the (UK) Ministry of Agriculture, Fisheries and Foods (MAFF), which acts as scientific secretariat to the Codex Committee on Fats and Oils; the Leatherhead Food RA, with the intention of developing and perfecting new analytical methodology; the Egyptian Government, whose interest was in cottonseed oil characteristics; and CPC (United Kingdom) Ltd, which had a particular interest in maize oil purity. MAFF was also interested in the nutritional aspects of the major vegetable oils with regard to assessment of the nation's diet.

The approach employed in the authenticity project was to obtain authentic commercial samples of vegetable oilseeds, to confirm the identity of the seeds and remove foreign material by visual inspection, and extract oil in the laboratory from these seeds of known origin and history. The oils were thus known to be authentic. In the case of palm oil, samples were obtained directly from plantation mill managers. Hand-picked experimental agricultural varieties and botanical specimens of seeds were deliberately excluded from this study, which related only to oils of commercial interest. The oils were then analysed by appropriate techniques such as fatty acid composition by GLC, fatty acids at the triglyceride 2-position by lipase hydrolysis and GLC, sterol composition by saponification of the oil followed by TLC and GLC, tocopherols content by HPLC, triglyceride carbon number by high-temperature GLC, stable carbon isotope ratio analysis (SCIRA) by burning the sample to CO_2, purifying the CO_2 by simple GLC and isotope ratio mass spectrometry. Other appropriate techniques such as iodine value melting point and solid fat content were determined, as the need arose. Samples were chosen from all the main geographical production areas.

The bulk of the traditional work took place in the period up to 1988, although there has been some updating since then, especially with regard to the newer varieties of rapeseed oil. In particular, we have been developing the new technique of stable carbon isotope ratio analysis as a criterion of purity in maize oil in the period 1989 to 1992.

Table 1 shows the number of countries of origin for each of the major oils, and the total number of samples studied up to 1988.

Table 1

Oil types studied in period 1981 to 1988

Oil type	No. of samples	No. of geographical origins
Palm oil and fractions	57	11
Groundnut	69	16
Sunflower-seed	50	14
Soya-bean	40	11
Rapeseed	91	16
Maize	36	10
Cottonseed	58	17
Palm kernel	79	20
Coconut	43	9
Safflower-seed	34	9
Babassu	5	2
Sesame	10	5
Total	572	

This work has been fully reported to MAFF, FOSFA International and Food RA Members in a series of reports describing the separate aspects of the work. We have now issued over thirty reports and papers on purity criteria in the series.[1-31]

It is appropriate to illustrate the breadth and scope of the work by reference to particular oils.

3. ANALYTICAL RESULTS AND DISCUSSION

Palm Oil

Table 2 shows the ranges and mean values of the fatty acid concentrations in Malaysian palm oil, and palm oil from all origins[4,5,18]. For comparison, figures have also been included for Malaysian palm oil obtained by the Palm Oil Research Institute of Malaysia (PORIM)[4] on Malaysian palm oil, the figures published by the Dutch National Standards Institute (NNI) in NPR 6305[32], and both the old and the recently revised (1993) Codex figures[33,34]. Some data for palm stearins and oleins, measured at the Food RA, are also shown[4,5]. The Food RA, PORIM and NNI figures agree quite closely, but the old Codex figures were much too divergent, and allowed impure oils to be passed off as good within the terms of the old Codex standard.

Sterol and tocopherol concentrations were also measured,[5] but these were not much help in judging whether a sample was a mixture containing palm fractions or a whole natural palm oil. The triglyceride carbon number composition was of some use here, as stearins had higher amounts of tripalmitin, of carbon number 48, while oleins had almost zero levels[4,5].

We found that the best technique, one initially suggested by PORIM[35], was to plot the iodine value (IV) against the slip melting (mpt), as shown in Figure 1. PORIM calculated regression lines relating slip melting point (y) with the iodine value (x) as shown on Figure 1.

Rapeseed and Soya-bean Oils

Again, fatty acid composition has been the first line of attack, and we have studied fatty acid compositions of rapeseed oil in some considerable depth, mainly because of the changes in agricultural varieties over the last ten years with the progressive shift from single-zero low-erucic acid strains to the double-zero crops, which are also low in glucosinolates. Table 3 shows the variations in RP composition according to variety[36]. Many of these samples were analysed after 1988 and are additional to those shown in Table 1.

In general, the different varieties all have similar fatty acid compositions, except that the linolenic (18:3)

TABLE 2

Fatty acid compositions of palm oil. Range (mean) (% m/m)

Fatty Acid	Food RA data [4,5,18]		PORIM data [4]	Dutch data [32]	Codex data [33,34]		Palm Stearin [4,5]	Palm Olein [4,5]
	All origins	Malaysian	Malaysian oils	NPR 6305	Pre 1987	Revised		
C12:0	ND - 0.2 (0.1)	ND - 0.1 (0.05)	0.1 - 1.0 (0.2)	0 - 0.1 (trace)	up to 1.2	up to 0.4	0.1 - 0.2	0.1 - 0.2
C14:0	0.8 - 1.3 (1.0)	0.9 - 1.1 (1.0)	0.9 - 1.5 (1.1)	0.8 - 1.3 (1.0)	0.5 - 5.9	0.5 - 2.0	1.0 - 1.3	0.9 - 1.0
C16:0	43.1 - 46.3 (44.3)	43.1 - 45.3 (44.1)	41.8 - 46.8 (44.0)	41 - 45 (43)	32 - 59	40.1† - 47.5	46.5 - 68.9	39.5 - 40.8
C16:1	tr - 0.3 (0.15)	0.1 - 0.3 (0.15)	0.1 - 0.3 (0.1)	0.1 - 0.3 (trace)	up to 0.6	up to 0.6	tr - 0.2	tr - 0.2
C18:0	4.0 - 5.5 (4.6)	4.0 - 4.8 (4.4)	4.2 - 5.1 (4.5)	4.5 - 6.0 (5.0)	1.5 - 8.0	3.5 - 6.0	4.4 - 5.5	3.9 - 4.4
C18:1	36.7 - 40.8 (38.7)	38.4 - 40.8 (39.6)	37.3 - 40.8 (39.2)	36 - 41 (38.5)	27 - 52	36 - 44	19.9 - 38.4	42.7 - 43.9
C18:2	9.4 - 11.9 (10.5)	9.4 - 11.1 (10.1)	9.1 - 11.0 (10.0)	10 - 12 (11.0)	5.0 - 14.0	6.5 - 12.0	4.1 - 9.3	10.6 - 11.4
C18:3	0.1 - 0.4 (0.3)	0.1 - 0.4 (0.2)	0.0 - 0.6 (0.4)	0.1 - 0.4 (trace)	up to 1.5	up to 0.5	0.1 - 0.2	ND - 0.4
C20:0	0.1 - 0.4 (0.3)	0.1 - 0.4 (0.2)	0.2 - 0.7 (0.4)	0.3 - 0.5 (0.5)	up to 1.0	up to 1.0	0.1 - 0.3	0.1 - 0.3
C20:1	ND - 0.3 (0.1)	-	-	tr - 0.4 (trace)	-	-	-	-

ND = not detected

tr = trace (unquantified level of less than 0.05%)

† = The incorrect value of 41.0 - 47.5 shown in Codex Alinorm 95/17 is due to a typographical error.

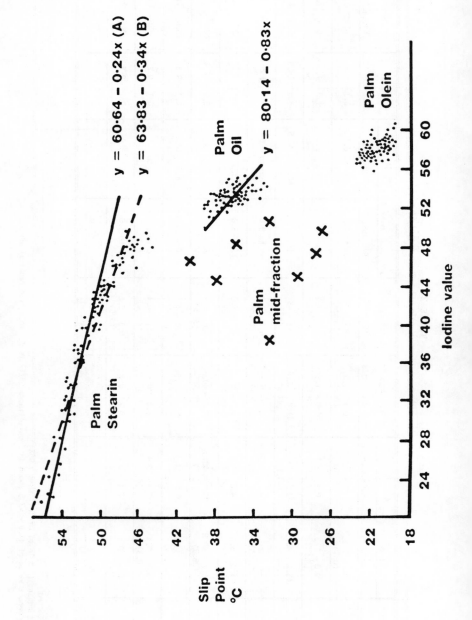

Figure 1   Slip point and iodine value

Table 3 Fatty acid ranges for rapeseed oils of known variety (weight %).[36]

Variety (No. of samples)	C14:0	C16:0	C16:1	C17:0	C17:1	C18:0	C18:1	C18:2	C18:3	C20:0	C20:1	C20:2	C22:0	C22:1	C24:0	C24:1
Ariana (5)	ND-0.1	4.5-5.3 (4.7)	0.2-0.3 (0.25)	ND-0.1	ND-0.1	1.5-1.7 (1.6)	57.9-60.0 (59.2)	21.0-22.0 (21.6)	9.4-11.0 (10.32)	0.5-0.6	0.9-1.2 (1.1)	ND-0.1	0.2-0.4	ND-0.1	ND-0.2	ND-0.2
Bienvenu (4)	ND	4.5-4.8 (4.7)	0.2-0.3	ND-0.1	ND-0.1	1.5-1.7 (1.6)	59.6-61.5 (60.5)	19.7-21.1 (20.6)	9.4-10.5 (9.9)	0.6	1.0-1.3 (1.15)	ND-0.1	0.3-0.4	ND-0.2	ND-0.1	ND-0.2
Ceres (2)	ND	4.8-5.0 (4.9)	0.2	ND	0.1	1.4-1.6 (1.5)	59.0-59.6 (59.3)	21.5-21.8 (21.65)	9.7-10.6 (10.15)	0.5-0.6	1.2	ND	0.3-0.4	ND	ND	ND
Cobra (2)	ND	4.9-5.0 (4.95)	0.2	ND	0.1	1.5	57.4-58.1 (57.75)	22.2-22.5 (22.35)	10.9-11.3 (11.1)	0.5-0.6	1.1-1.2	ND	0.3	ND-0.2	ND	ND
Drakkar (2)	ND	3.4-3.9 (3.65)	0.1-0.2	0.1-0.3	0.1-0.2	1.7-2.1 (1.9)	64.6-66.9 (65.75)	16.1-17.0 (16.55)	9.0-9.3 (9.15)	0.5-0.6	1.3-1.6 (1.45)	ND	0.3-0.4	ND-0.2	ND	ND
Global (2)	ND	4.0-4.2 (4.1)	0.2	ND-0.1	0.1	1.5-1.6 (1.55)	59.1-59.8 (59.45)	20.8-21.8 (21.3)	10.5-11.0 (10.75)	0.5-0.6	1.3-1.4 (1.35)	0.1	0.3	ND	ND	ND
Jet Neuf (2)	ND	4.7-4.8 (4.75)	0.2	ND-0.1	0.1	1.5-1.6 (1.55)	57.6-60.5 (59.05)	20.9-22.1 (21.5)	8.6-9.2 (8.9)	0.5	1.2-2.0 (1.6)	ND-0.1	0.3	ND-2.7 (1.35)	ND	ND
Libravo (2)	ND	4.1-4.2 (4.15)	0.2	ND-0.1	ND-0.1	1.5	60.5-63.0 (61.75)	19.8-20.9 (21.35)	9.0-10.4 (9.7)	0.5	1.3	ND-0.1	0.3	ND	ND-0.1	ND-0.1
Mikado (4)	ND	4.3-4.6 (4.4)	0.2-0.3	ND-0.1	ND-0.1	1.6-1.9 (1.8)	57.7-60.1 (58.95)	21.8-23.4 (22.55)	8.9-10.1 (9.4)	0.6-0.8	1.0-1.2 (1.1)	ND-0.1	0.4-0.5	ND-0.1	ND-0.2	ND-0.2
Pasha (4)	ND-0.1	4.5-4.6 (4.55)	0.2	ND	ND-0.2	1.6-1.8 (1.7)	58.7-63.9 (62.05)	18.1-21.1 (19.2)	9.0-11.1 (9.8)	0.6	1.2-1.4 (1.3)	ND-0.1	0.3	ND-0.2	ND-0.1	ND-0.2
Rafal (3)	ND-0.1	5.1-5.2 (5.1)	0.3	ND-0.1	ND-0.2	1.4-1.5 (1.4)	58.7-60.4 (59.6)	21.1-22.0 (21.6)	9.0-9.8 (9.3)	0.6	1.0-1.3 (1.1)	0.1	0.3-0.4	ND	ND-0.2	ND-0.2
Tapidor (2)	ND	4.5-4.9 (4.7)	0.2	ND	0.1-0.2	1.6-1.7 (1.65)	59.9-60.7 (60.3)	20.4-20.6 (20.5)	9.6-10.4 (10.0)	0.6	1.2-1.4 (1.3)	ND	0.3-0.4	ND-0.5	ND	ND
Topas (3)	ND-0.1	3.5-4.1 (3.8)	0.2-0.3	ND-0.1	ND-0.3	1.5-1.7 (1.6)	57.0-64.0 (60.0)	16.4-19.6 (18.5)	9.9-11.2 (10.8)	0.5-0.6	2.0-3.0 (2.4)	0.1	0.3-0.4	1.1-1.9 (1.4)	ND-0.1	ND-0.3

Table continues...

TABLE 3 Cont'd...

Single samples of the following varieties were analysed:-

Variety (No. of samples)	C14:0	C16:0	C16:1	C17:0	C17:1	C18:0	C18:1	C18:2	C18:3	C20:0	C20:1	C20:2	C22:0	C22:1	C24:0	C24:1
Cascade	ND	4.4	0.2	0.1	0.1	1.8	61.7	20.0	9.6	0.6	1.2	ND	0.3	ND	ND	ND
Darmor	ND	4.3	0.2	ND	0.1	1.5	58.8	22.0	10.8	0.5	1.3	ND	0.3	0.2	ND	ND
Diadem	ND	4.1	0.2	0.1	0.1	1.3	59.5	21.5	11.2	0.4	1.2	0.1	0.2	ND	ND	ND
Doublol	ND	3.8	0.2	ND	0.2	1.6	63.0	18.5	10.4	0.6	1.4	ND	0.3	ND	ND	ND
Lictor	ND	4.6	0.2	ND	0.1	1.5	60.3	21.0	10.1	0.5	1.3	ND	0.3	ND	ND	ND
Liradona	0.1	4.3	0.3	ND	ND	1.7	62.6	19.2	9.2	0.6	1.3	0.1	0.4	ND	0.1	0.1
Lirawel	ND	4.1	0.2	ND	ND	1.5	61.2	20.3	10.1	0.5	1.4	0.1	0.3	ND	0.1	0.2
Puma	ND	3.8	0.2	ND	0.1	1.5	60.4	21.1	10.6	0.5	1.4	0.1	0.3	ND	ND	ND
Score	ND	4.3	0.2	ND	0.1	1.5	61.2	20.6	9.9	0.5	1.3	0.1	0.3	ND	ND	ND
Sema	ND	4.3	0.2	0.1	0.1	1.5	61.4	20.3	10.0	0.6	1.2	ND	0.3	ND	ND	ND
Suzanna	ND	4.5	0.2	ND	0.1	1.8	61.5	21.0	8.8	0.6	1.1	ND	0.3	ND	ND	ND

ND = not detected

Mean values were not calculated for minor acids

Where only one result is given for a range, all samples had the same value, within experimental error.

level has crept up progressively from the 8-10% norm in former years to 14% in some of the newer rapeseed varieties. This must in due course lead to an inferior oxidative and/or flavour stability, unless these high linolenic acid varieties can be discouraged by the trade.

Table 4 shows the variation in soya-bean oil fatty acid compositions according to the country of origin[18].

On the whole, rapeseed oil contains 52-65% oleic acid (C18:1) while soyabean oil contains 50-57% linoleic acid (C18:2); conversely, rapeseed oil contains 17-25% linoleic, while soyabean oil contains 17-25% oleic. Distortion of these ranges is therefore good evidence for co-mingling. Both oils contain moderate levels of linolenic acid (C18:3).

Another indicator of rapeseed oil in soyabean oil is, of course, the brassicasterol content of rapeseed oil, pure soyabean oil containing none,[8, 9, 18, 25, 26, 31, 32, 36, 37].

Vegetable oil triglycerides do, of course, have fatty acids at the 2-, or central position with a different composition from those at the 1-, and 3- positions. This is particularly so with cocoa butter and it is used as a purity criterion with olive oil[38]. We found that there was a constant ratio between the concentration of an acid at the 2-position and its concentration overall, taken over a range of oils all of the same variety. What is more, this ratio varies from oil to oil. This is shown in Figure 2, where the concentration of C18:3 at the triglyceride 2-positions in both rapeseed oil and soyabean oil are plotted against the overall concentration to give the Enrichment Factor (EF) for the relevant acid[37]. We have found this a useful criterion in substantiating views deduced from other tests.

This difference in C18:3 concentrations at the triglyceride 2-position will, of course, result in a different triglyceride composition, and could lead to a purity criterion based on HPLC analysis of these triglycerides. However, the present technique is more easily applicable in less sophisticated laboratories. Furthermore, Figure 2 illustrates that levels of C18:3 at the 2-position do not overlap in the two oils.

Tocopherol Concentrations in Vegetable Oils

The compositions of tocopherols in oils also vary in a characteristic way. Tocopherols are of course, anti-oxidants and their levels will fall as an oil ages. However, the ratios of different tocopherol concentrations stay more nearly constant. What is more, tocopherol levels cannot increase as an oil ages. Table 5 shows the concentrations of tocopherols and tocotrienols measured in various oils at the Food RA[13].

Most oils contain zero levels of tocotrienols, with the major exception of palm oil. Oil from whole maize does

Table 4 Fatty acid compositions for soya-bean oil according to origin[18]

Fatty acid compositions (wt % methyl esters)

Origin (No. of samples)	C16:0	C16:1	C18:0	C18:1	C18:2	C18:3	C20:0	C20:1	C22:0	C22:1	C24:0
Argentina (6)	10.7-11.0	tr-0.2	3.7-4.3	17.7-19.1	55.0-56.8	7.9-9.5	0.2-0.5	0.2*	0.5-0.6	tr-0.2	ND-0.2
Brazil (6)	10.7-11.8	tr-0.2	3.7-4.2	21.9-23.9	53.1-54.7	5.9-7.3	0.4-0.5	0.2-0.3	0.5-0.7	tr-0.3	tr-0.2
Paraguay (4)	11.3-12.2	tr-0.2	3.7-3.9	22.5-22.7	54.0-54.4	5.6-6.4	0.3-0.5	0.2*	0.5-0.6	0.2-0.3	0.2*
Uruguay (2)	11.4-11.7	tr-0.1	4.2*	19.8-20.4	53.7-54.6	8.3-8.8	0.4*	0.2*	0.4-0.5	tr-0.1	0.2*
USA (8)	10.4-11.9	0.1-0.2	3.9-4.8	22.3-25.5	50.5-53.6	5.5-8.1	0.3-0.6	0.2-0.3	0.3-0.6	ND-0.1	ND-0.4
Canada (3)	10.8-10.9	0.1*	3.6-4.0	21.9-24.7	51.2-52.9	8.4-9.3	0.3	0.2*	0.3-0.4	ND-tr	0.1*
S Africa (3)	9.9-10.7	0.1-0.2	4.0-5.4	18.8-21.9	53.2-55.8	7.9-9.1	0.4-0.5	0.2*	0.4-0.5	tr*	0.1-0.2
Australia (2)	11.1-11.7	tr	3.4-3.8	21.7-21.9	54.2-54.8	7.5-8.0	0.2-0.3	tr-0.1	0.4*	ND-0.1	0.1*
India (3)	9.7-11.7	tr-0.2	3.5-4.3	22.4-26.1	49.8-54.7	6.8-7.1	0.3-0.4	0.2-0.3	0.4-0.5	0.1-0.2	0.1-0.2
France (1)	10.9	tr	3.0	19.8	57.1	8.4	0.1	0.1	0.3	ND	0.1
China (1)	13.3	tr	4.0	20.4	51.9	9.1	0.5	0.3	0.4	ND	ND
Overall range (39)	9.7-13.3	tr-0.2	3.0-5.4	17.7-26.1	49.8-57.1	5.5-9.5	0.1-0.6	tr-0.3	0.3-0.7	ND-0.3	ND-0.4
Mean (39)	11.2	0.1	4.0	21.8	53.9	7.6	0.4	0.2	0.5	0.1	0.1
Coefficient of variation % (39)	6.0	68.4	9.9	9.3	2.9	15.7	26.1	25.6	19.2	124.6	61.1
Codex range (1)	8.0-13.3	0-0.2	2.4-5.4	17.7-26.1	49.8-57.1	5.5-9.5	0.1-0.6	0-0.3	0.3-0.7	0-0.3	0-0.4

Note: (1) Codex Standard as revised September 1993[34]

* Where single values are given and more than one sample was analysed, all samples had the same value within experimental error.

tr = trace (unquantified level of less than 0.05%)

ND = not detected

Table 5 Ranges (means) of tocopherol and tocotrienol levels in vegetable oils[13]
(mg/kg)

Tocol	Palm kernel--	Coconut*	Cottonseed	Soya-bean	Maize--	Groundnut	Palm	Sunflower-seed	High-erucic rapeseed	Low-erucic rapeseed
					Oils analysed					
aT	ND-44	ND-17	136-543 (388)	9-252 (99.5)	23-573 (282)	49-304 (170)	4-105 (89)	403-855 (670)	39-305	100-320 (202)
ßT	ND-248	ND-11	ND-29 (16.9)	ND-36 (7.7)	ND-356 (54)	0-41 (8.8)	-	9-45 (27)	24-158	16-140 (65)
yT	ND-257	ND-14	158-594 (429)	409-2,397 (1,021)	268-2,468 (1,034)	99-389 (213)	6-36 (18)	ND-34 (11)	230-500	287-753 (490)
δT	-	ND-2	ND-17 (3.3)	154-932 (421)	23-75 (54)	3-22 (7.6)	-	ND-7 (0.6)	5-14	4-22 (9)
aT3	ND-tr	ND-5	-	-	ND-239 (49)	-	4-336 (120)	-	-	-
ßT3	-	-	-	-	ND-52 (0)	-	-	-	-	-
yT3	ND-60	ND-1	-	-	ND-450 (161)	-	42-710 (323)	-	-	-
δT3	-	-	-	-	ND-20 (6)	-	tr-148 (72)	-	-	-
Total	ND-257	tr-31	410-1,169 (788)	575-3,320 (1,549)	331-3,402 (1,647)	176-696 (407,4)	90-1,327 (630)	447-900 (709)	312-928	424-1,054 (766)

- Means are negligible.
- High values may be due to migration of some palm oil into the palm kernels before separation.
Means are shown below ranges.
-- Oil obtained from maize germ contains no tocotrienols.

T = totopherol   T3 = tocotrienol

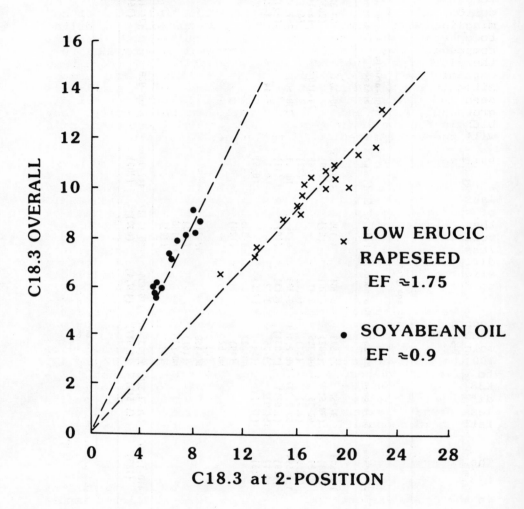

Figure 2   Plot of C18:3 overall vs C18:3 at the 2-position

contain tocotrienols but commercial oil, that is oil obtained from the separated maize germ, is free of tocotrienols. The presence of tocotrienols in, e.g. cottonseed oil is therefore an immediate indicator of co-mingling with palm. Groundnut oil contains no delta-tocopherol, while soya-bean oil is quite rich in this compound. The presence of δ-tocopherol in groundnut oil therefore substantiates any allegation of soya-bean contamination deduced from, say, the fatty acid compo-sition. The mean ratio of α- to γ-tocopherol in sunflower-seed oil is over 200, while other oils such as cotton, groundnut, maize and soya have ratios of below unity. Any blending of sunflower-seed oil with one of these other oils will therefore immediately influence this ratio.

Palm Kernel and Coconut Oils

Palm kernel (PK) and coconut oils (CN) are rich in lauric acid, and while coconut oil contains higher levels of the acids with 6, 8 and 10 carbons, the methyl esters of these acids can be easily lost, owing to their higher volatility, leading to occasional doubts about analytical results and/or published ranges. In any case, the differences between palm kernel oil and coconut oil are slight and low levels of one oil in the other are difficult to detect by reference to fatty acid compositions alone.

We found that carbon number analysis is a very good technique with the lauric oils. We recalculate the carbon number composition to exclude most diglycerides and the molecules of higher molecular weight by re- normalising to 100 the triglycerides with carbon numbers between 32 and 42, to give "K" values, and plotting K34 + K40 against K36 + K38, as shown in Figure 3. This gives a clear differentiation of palm kernel oil and coconut oil, and is less prone to experimental error than the determination of fatty acid compositions[11,12].

The Codex Alimentarius Fats and Oils Specifications

As mentioned previously, the fatty acid compositions in the Codex Alimentarius were too wide, and allowed impure oils to pass as pure. This is not surprising as the Codex ranges were derived from a survey of literature data, some of which were very old and of unknown reliability. Furthermore, agricultural practices have changed, leading to variations over the years of the compositions of some vegetable oils.

We presented our data to the Codex Fats and Oils Committee, and when these were taken in conjunction with data from some other institutes, a worthwhile revision and 'tightening up' of the Codex specifications was achieved in 1987. This revision of Codex data continued and a further revision[34] was effected in October 1993 by taking account of the data that the Food RA had updated since the 1987

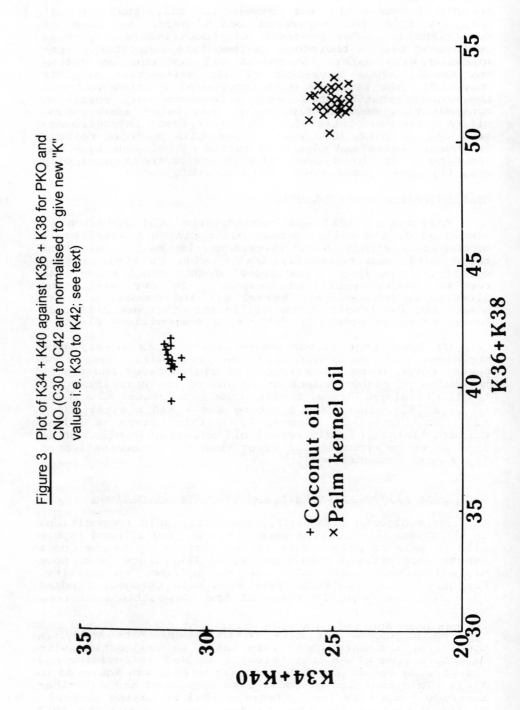

Figure 3   Plot of K34 + K40 against K36 + K38 for PKO and CNO (C30 to C42 are normalised to give new "K" values i.e. K30 to K42; see text)

revision. The main value of the Food RA data was that the origin and history of each sample were known, and the range of samples analysed covered all the main harvest regions. The fatty acid composition ranges in the proposed draft Codex standard for Named Vegetable Oils are now very similar to those values obtained at the Leatherhead Food RA. Revised Codex ranges are shown in Tables 2, 6 and 7. These are now at step 4 of the 8 step Codex Procedure for the elaboration of standards. The proposed draft standard for Named Vegetable Oils containing these FAC will be discussed at the next Codex Alimentarius Commission meeting in 1995. The revised values shown in Tables 2, 6 and 7 do not therefore, at the time of writing, yet form part of the final standards.

Maize Oil Purity by Stable Carbon Isotope Ratio Analysis

Table 7 shows figures for the fatty acid composition of maize oil determined at the Leatherhead Food RA[20], alongside the old Codex ranges, and the 1993 Codex ranges adopted following consideration of the information provided by the Leatherhead Food RA. The Food RA study also established ranges for tocopherol, sterol, tri- glyceride and 2-position fatty acid compositions, as mentioned previously.

Unfortunately, establishing maize oil purity is no simple task, as the fatty acid compositions of several other liquid oils overlap with that of maize oil, and in some respects lie on either side of it. It is therefore possible to form blends of other oils, which appear very similar to maize oil in this respect. In addition, maize oil contains more sterols than any other oil, with a consequence that the sterol composition of any blend is swamped by the maize component in the blend. The purity criteria of maize oil is therefore a matter of concern, and a new method of establishing maize oil purity was a desirable objective.

Our attention was drawn to the technique of isotopic ratio analysis, which has been used to detect impurity in various types of sugar syrups[39,40,41], vanillin[42] and apple juice[43,44]. This technique now shows promise of being used as a new purity criterion for maize oil.

Isotopic ratio analysis involves burning the sample to form carbon dioxide, purifying this by GLC to remove other gaseous oxides, and then analysing the relative proportions of carbon 12 and carbon 13 isotopes of carbon by a sensitive mass spectrometer. This technique is quite straightforward. It has been applied to a wide variety of foods. It has been well described elsewhere [40,41,42,43,44,45,46,47,48,49,50], and with modern equipment is no more expensive than the determination of fatty acid composition. It is a technique well established in the mineral oil industry, where it is used for investigating the nature and quality of oil from oil wells in different parts of the world. The isotopic ratio is defined by the isotopic enrichment factor, as shown

Table 6 Codex Alimentarius fatty acid ranges (Revised 1993)[34]

(% of total fatty acids)

(Ranges refer to commercial samples of bona fide fats and oils)

Fatty acid	Groundnut	Safflower	Soya	Sunflower	Low erucic acid rape	"Normal" rape	Coconut	Palm kernel
C6:0	ND	ND	ND	ND	ND	up to 0.1	0.0 - 0.6	up to 0.8
C8:0	ND	ND	ND	ND	ND	up to 0.1	4.6 - 9.4	2.4 - 6.2
C10:0	ND	ND	ND	ND	ND	up to 0.1	5.5 - 7.8	2.6 - 5.0
C12:0	up to 0.1	ND	0.0 - 0.1	0.0 - 0.1	ND	up to 0.1	45.1 - 50.3	41 - 55
C14:0	up to 0.1	0 - 0.2	0.0 - 0.2	0.0 - 0.2	up to 0.2	up to 0.2	16.8 - 20.6	14 - 18
C16:0	8.3 - 14.0	5.3 - 8.0	8.0 - 13.3	5.6 - 7.6	3.3 - 6.0	1.5 - 6.0	7.7 - 10.2	6.5 - 10
C16:1	up to 0.2	0.0 - 0.2	0.0 - 0.2	0.0 - 0.3	0.1 - 0.6	up to 3.0	ND	NS
C18:0	1.9 - 4.4	1.9 - 2.9	2.4 - 5.4	2.7 - 6.5	1.1 - 2.5	0.5 - 3.1	2.3 - 3.5	1.3 - 3.0
C18:1	36.4 - 67.1	8.4 - 21.3	17.7 - 26.1	14 - 39.4	52.0 - 66.9	8 - 60	5.4 - 8.1	12 - 19
C18:2	14.0 - 43.0	67.8 - 83.2	49.8 - 57.1	48.3 - 74.0	16.1 - 24.8	11 - 23	1.0 - 2.1	1.0 - 3.5
C18:3	up to 0.1	0.0 - 0.1	5.5 - 9.5	0.0 - 0.2	6.4 - 14.1	5 - 13	0.0 - 0.2	up to 0.1
C20:0	1.0 - 1.7	0.2 - 0.4	0.1 - 0.6	0.2 - 0.4	0.2 - 0.8	up to 3.0	0.0 - 0.2	up to 0.1
C20:1	0.7 - 1.7	0.1 - 0.3	0.0 - 0.3	0.0 - 0.2	0.1 - 3.4	3 - 15	0.0 - 0.2	up to 0.1
C22:0	2.1 - 4.4	0.2 - 0.8	0.3 - 0.7	0.5 - 1.3	0 - 0.5	up to 2.0	ND	up to 0.1
C22:1	up to 0.3	0.0 - 1.8	0.0 - 0.3	0.0 - 0.2	0.0 - 2.0	5 - 60	ND	up to 0.1
C24:0	1.1 - 2.2	0.0 - 0.2	up to 0.4	0.0 - 0.3	up to 0.2	up to 2.0	ND	up to 0.1
C24:1	0 - 0.3	0.0 - 0.2	ND	0.2 - 0.3	up to 0.4	up to 3.0	ND	up to 0.1

ND = not detected　　　NS = not specified

in Figure 4.

The isotopic ratio can be influenced[40,43,45,50] by i) the equilibrium between atmospheric carbon dioxide and the carbon dioxide dissolved in water (e.g. sea water) the carbon dioxide in water being enriched with carbon 13; and ii) photosynthesis in plants. The latter aspect is of most importance in the current project. The photosynthetic route of carbon dioxide fixation differs from plant to plant, there being three main plant groupings. These are differentiated by the number of carbons in the major initial metabolite, the C_3 route being to phospho-glyceric acid, the C_4 to oxaloacetic acid, and the CAM or Crassulacean Acid Metabolism (CAM) route. The isotopic enrichment factor usually lies in the range −24 to −30 for the C_3 route, −9 to −14 for the C_4 route, and −12 to −30 for the CAM route. The main interest in the present project is that maize is a C_4 plant, whereas all the other major vegetable oils are obtained from C_3 plants.

Table 7

Fatty acid composition ranges for maize oil

| | Food RA | Codex | |
	Range	Old	Revised*
C12:0	0−0.3	0−0.1	0−0.3
C14:0	0−0.3	0−0.1	0−0.3
C16:0	10.7−16.5	8.0−19	8.6−16.5
C16:1	0−0.4	0−0.5	0−0.4
C18:0	1.6−3.3	0.5−4.0	1.0−3.3
C18:1	24.6−42.2	19−50	20−42.2
C18:2	39.4−60.4	34−62	39.4−62.5
C18:3	0.7−1.3	0−2.0	0.5−1.5
C20:0	0.3−0.6	0−1.0	0.3−0.6
C20:1	0.2−0.4	0−0.5	0.2−0.4
C22:0	0−0.5	0−0.5	0−0.5
C24:0	0.1−0.4	0−0.5	0−0.4

*Revised at Codex Alimentarius Fats and Oils Committee October 1993[34].

$$\delta^{13}C = \left\{ \frac{[^{13}C/^{12}C] \text{ sample} - [^{13}C/^{12}C] \text{ standard}}{[^{13}C/^{12}C] \text{ standard}} \right\} \times 10^3$$

Mean abundances in air ^{12}C − 98.9%
^{13}C − 1.1%

$[^{13}C/^{12}C]$ standards(NBS22 lubricating oil = −29.8‰ vs PDB)
(Pee Dee Belemnite (Carolina)

Figure 4   Isotopic ratio analysis

Therefore, a number of new samples of authentic maize seed and separated maize germ were obtained from a wide variety of geographical origins. Foreign materials were manually removed from the seeds or germ, and the oil was extracted in the laboratory. The fatty acid compositions and the isotopic ratios of the extracted oils were determined[42]. This was done in comparison with a number of new samples of the other major vegetable and animal fats. Table 8 shows the origins of the new maize samples. Table 9 shows the isotopic ratio results for the non-maize vegetable oils, while Table 10 shows maize oil results alongside selected non-maize values.

At this stage of the work, it became apparent that the isotope ratio mass spectrometer was not giving accurate results in the -10 to -20 range[51], apparently because the MS machine used had drifted out of linearity. It had been regularly calibrated with an NSB22 lubricating oil standard with a carbon isotope ratio of -29.8[52], and the non-maize results were shown to be reliable. Extensive work[53], involving comparison of results between three different laboratories, with coded samples, some of which were blind duplicates, established which machines gave the results of greatest reliability[53]. All the maize results given in the present paper were checked carefully in this exercise, and we now believe them to be reliable. Earlier results[52] may in some cases be of poor accuracy.

It can be seen that the results for maize range from -13.71 to -16.36 with a mean of -14.95. The non-maize results are clearly differentiated, ranging from -32.39 to -25.38. Two sesame seed oil results are extreme and were checked several times the reported values being always confirmed, while the cereal oils are not serious candidates for corn oil contamination. However, the extreme values for the oils from wheat, barley and oat germs may enable identification of contamination in these high-value products themselves.

Table 8

Origins of New Maize Samples

Argentina	11	New Zealand	1
Australia	3	Nigeria	2
Brazil	2	Spain	2
France	4	Turkey	2
Greece	3	USA	7
Italy	2	Other origins	3

As mentioned previously, the $^{12}C/^{13}C$ ratio of CO_2 in water is different from that in air, so the composition of fish, which feed ultimately on plankton, etc., may also be expected to vary. This is shown in the results, as the fish oil isotopic ratios are less negative than the other non-maize results.

Table 9

Isotope Ratio Results of some Vegetable Oil Samples (Non-maize)

Oil type (No. of samples)	$\delta^{13}C$ Range (mean)
Cottonseed (8)	−27.40 to −28.28 (−27.78)
Sunflowerseed (5)	−27.94 to −29.76 (−28.95)
Rapeseed (7)	−27.47 to −29.40 (−28.56)
Safflowerseed (6)	−27.87 to −30.17 (−28.94)
Soyabean (6)	−29.67 to −30.55 (−30.09)
Sesameseed (7)	−25.38 to −29.28 (−27.93)
Groundnut kernel (7)	−26.48 to −28.28 (−27.78)
Palm olein (6)	−29.51 to −29.84 (−29.65)
Palm oil (6)	−29.25 to −29.91 (−29.64)
Palm kernel (10)	−27.49 to −30.27 (−29.47)
Others (5)*	−28.90 to −32.39 (−30.79)
Non-maize vegetable results (73)	−25.38 to −32.39 (−28.99)

* Wheat, barley, oat germ oils, rice bran oil and virgin olive oil.

Table 10

Comparison of isotopic ratio results

Oil	Mean	Range
Maize (42)	−14.95	−13.71 to −16.36
Non-maize vegetable (73)	−28.99	−25.38 to −32.39
Fish oils (4)	−26.66	−25.37 to −27.95
Animal fats (5)	−30.28	−27.56 to −32.08

Animals are sometimes fed on corn or maize fodder and the query therefore arises whether animal fats may reflect the isotopic ratio of maize. So far, none of the isotopic ratios measured for animal fats shows this trend. However, fat samples obtained from beef cattle or chicken fed exclusively on maize for the whole of their lives are still being sought.

A number samples taken from the same maize oil source have also been examined in order to establish repeatability and reproducibility of the technique (Table 11).

The results so far obtained show considerable promise for this new method of establishing maize oil purity.

The question then arises of how low a degree of contamination can be detected by this new technique. Bearing in mind that fish oils and animal fats can be easily excluded by other means, and that cereal oils are unlikely contaminants, and together with the natural ranges of the isotopic ratios found, our present estimate is that levels of 10% foreign oil in a maize sample can always be established. For a maize oil sample lying in the middle of the isotopic ratio range, contamination with less than 10% of an average non-maize vegetable oil can be easily detected.

To some extent this reflects the limits of detection of contaminants in other (non-maize) oils by conventional means such as fatty acid composition. A study of statistical pattern recognition techniques, and canonical variance is now seen as the next stage. By thus combining the two approaches of isotopic ratio and fatty acid composition it is hoped that the sensitivity could be doubled thus enabling detection of 5%, or better, of vegetable oil contamination in maize oil. An especially

Table 11

Isotope ratio replication

Groundnut (6 samples)		Maize oil (single sample)		
−28.44,	−28.47	−15.23		
−28.04,	−28.13	−15.01,	−15.27*	
−28.68,	−28.70	−15.40		
−27.24,	−27.17	−15.34		
−28.63,	−28.63	−15.34		
−27.85,	−27.92	−15.39		
		−15.21		
		−15.35		
		−15.28		
		−15.30		
		−15.35,	−15.31,	−15.48*
		−15.37		
		−15.40		
		−15.35,	−15.36	
		−15.39		
Range −27.17 to −28.70		−15.01 to −15.48		
Mean −28.16		−15.31		
S.D.+ 0.53		0.098		

* Duplicate and triplicate results determined at same time.
+ Standard deviation

attractive technique currently under evaluation combines data on iodine value and SCIR, which appears capable of detecting levels of contaminant as low as 4%. One problem remains - that of ensuring that any experimental error, such as the non-linearity of the mass spectrometer used to determine the isotope ratio noted earlier, does not so distort the work that unreliable conclusions are drawn. This is routinely guarded against by always analysing a sub-sample of the bulk stock of maize oil used for the data in Table 11, and for which fully reliable data on the acceptable SCIR are available.

4. ACKNOWLEDGEMENT

The author wishes to express his thanks to the Management at the Leatherhead Food RA, The Ministry of Agriculture Fisheries and Foods, CPC, and FOSFA International for permission to give this paper, to Mr M J Downes, Mr P N Gillatt and Miss I A Bentley for technical assistance, and to Mrs A Pernet for editorial help with the text.

5. REFERENCES

1. A.A. Abdel-Nabey, A. Adel-Shehata, S.P. Kochhar and J.B. Rossell, Lebensmittel Wiss U. Technol., 1986, 19, 393-4.

2. J.A. Turrell, B. King, S.A. Zilke and A. Abdel-Nabey "Authenticity of cottonseed oil." Leatherhead Fd RA Res.Rep. No. 607, 1987.[†]

3. B. King "Authenticity of oils and fats. Part I General Introduction". Leatherhead Fd RA Res.Rep. No. 438, 1983.[†]

4. J. B. Rossell, B. King and M.J. Downes J.Am.Oil Chem Soc. 1985, 62, 221-30.

5. B. King and I. Sibley "Authenticity of edible vege-table oils and fats. Part II. Palm oil and palm oil fractions." Leatherhead Fd RA Res. Rep. No.462,1984.[†]

6. B. King and I. Sibley "Authenticity of edible vegetable oils and fats. Part III. Groundnut oil." Leatherhead Fd RA Res. Rep. No. 494, 1984.[†]

7. B. King, I. Sibley and S.A. Zilka "Authenticity of edible vegetable oils and fats. Part IV. Sunflower-seed oil." Leatherhead Fd RA Res. Rep. No. 520, 1985.[†]

8. B. King, I. Sibley and S.A. Zilka "Authenticity of edible vegetable oils and fats. Part V. Soya-bean oil." Leatherhead Fd RA Res. Rep. No. 464, 1984.[†]

9. B. King, I. Sibley and S.A. Zilka "Authenticity of edible vegetable oils and fats. Part VI Rapeseed oil." Leatherhead Fd RA Res. Rep. No. 515, 1985.[†]

10. B. King, I. Sibley and S.A. Zilka "Authenticity of edible vegetable oils and fats. Part VII Maize oil."

Leatherhead Fd RA Res. Rep. No. 522, 1985.

11.   B. King and S.A. Zilka "Authenticity of edible
       vegetable oils and fats. Part IX. Palm kernel oil."
       Leatherhead Fd RA Res. Rep. No. 559, 1986.[†]

12.   B. King, S.A. Zilka and J.A. Turrell "Authenticity of
       edible oils and fats. Part X. Coconut oil." Leatherhead
       Fd RA Res. Rep. No. 531, 1985.[†]

13.   B. King, J.A. Turrell and S.A. Zilka "Authenticity of
       edible vegetable oils and fats. Part XI. Analysis of
       minor fatty acid components by capillary column GLC,
       and of triglycerides by HPLC." Leatherhead Fd RA Res.
       Rep. No. 563, 1986.[†]

14.   J.A. Turrell, P.A. Whitehead and S.A. Zilka
       "Authenticity of edible vegetable oils and fats. Part
       XII. Safflower-seed oil." Leatherhead Fd RA Res. Rep.
       No. 599, 1987.[†]

15.   J.A. Turrell, P.A. Whitehead and P.M. Dagnall
       "Authenticity of edible vegetable oils and fats. Part
       XIII Substituted sterols." Leatherhead Fd RA Res. Rep.
       No. 623, 1988.[†]

16.   J.A. Turrell, P.A. Whitehead and D.M. Rose
       "Authenticity of edible vegetable oils and fats. Part
       XIV. Statistical evaluation of vegetable oil purity
       criteria." Leatherhead Fd RA Res. Rep. No. 614, 1988.[†]

17.   J.A. Turrell and P.A. Whitehead "Authenticity of edible
       vegetable oils and fats. Part XV. Supplementary report
       on coconut, palm kernel and babassu kernel oils."
       Leatherhead Fd RA Res. Rep. No. 634, 1989.[†]

18.   J.A. Turrell and P.A. Whitehead "Authenticity of
       edible vegetable oils and fats. Part XVI. Analysis of
       additional samples of palm, soyabean and rapeseed
       oils." Leatherhead Fd RA Res. Rep. No. 665,
       April. 1990[†]

19.   J.A. Turrell and P.A. Whitehead "Authenticity of edible
       vegetable oils and fats. Part XVII. Analysis of
       triglycerides by high-performance liquid
       chromatography." Leatherhead Fd RA Res. Rep. No. 640,
       1989.[†]

20.   J.A. Turrell and P.A. Whitehead "Authenticity of edible
       vegetable oils and fats. Part XVIII. Analysis of
       additional samples of sunflower-seed, groundnut and
       maize germ oils." Leatherhead Fd RA Res. Rep. No. 637,
       1989.[†]

21.   M.J. Downes "Determination of the sterol composition
       of crude vegetable oils Part 1. General introduction
       and method of analysis." Leatherhead Fd RA Tech. Circ.
       No. 780, 1982.[†]

22. M.J. Downes "Determination of the sterol composition of crude vegetable oils. Part II. Palm oil and palm fractions." Leatherhead Fd RA Tech. Circ. No. 781, 1982.[†]

23. M.J. Downes "Determination of the sterol composition of crude vegetable oils. Part III. Groundnut oil." Leatherhead Fd RA Res. Rep. No. 407, 1983.[†]

24. M.J. Downes "Determination of the sterol composition of crude vegetable oils. Part IV. Sunflower-seed oil." Leatherhead Fd RA Res. Rep. No. 455, 1984.[†]

25. M.J. Downes "Determination of the sterol composition of crude vegetable oils. Part V. Soya-bean oil." Leatherhead Fd RA Res. Rep. No. 487, 1984.[†]

26. M.J. Downes "Determination of the sterol composition of crude vegetable oils. Part VI. Rapeseed oil." Leatherhead Fd RA Res. Rep. No. 436, 1983.[†]

27. M.J. Downes "Determination of the sterol composition of crude vegetable oils. Part VII. Maize oil." Leatherhead Fd RA Res. Rep. No. 441, 1983.[†]

28. M.J. Downes "Determination of the sterol composition of crude vegetable oils. Part VIII. Cottonseed oil." Leatherhead Fd RA Res. Rep. No. 518, 1985.[†]

29. M.J. Downes "Determination of the sterol composition of crude vegetable oils. Part IX. Palm kernel oil." Leatherhead Fd RA Res. Rep. No. 519, 1985.[†]

30. M.J. Downes "Determination of the sterol composition of crude vegetable oils. Part X. Coconut oil." Leatherhead Fd RA Res. Rep. No. 516, 1985.[†]

31. S.P. Kochlar "The effect of processing on vegetable oil sterols - A literature review." Leatherhead Fd RA Sci. Tech. Surv. No. 133, 1982.[†]

32. Dutch National Standards Institute (NNI). NPR 6305 January 1984.

33. Codex Alimentarius Alinorm 79/17.

34. Codex Alimentarius Alinorn 95/17.

35. PORIM "Technology", No. 4, May 1981, p.6.

36. M.J. Downes, J. Sollers, M.A. Jordan and P.A. Whitehead "Authenticity of edible vegetable oils and fats, Part XIX." "Analysis of Additional Samples of Rapeseed Oil, 1986-88 Crop Years" Leatherhead Fd RA Res. Rep. No. 682, 1990.[†]

37. J.B. Rossell "Soya-bean and rapeseed oils." Leatherhead Fd RA Symposium proc. No. 31, p.81, 1985.[†]

38. Commission Regulation (EEC) No. 2568/91 of 11 July 1991
     on the characteristics of olive oil and olive residue
     oil and on the relevant methods of analysis. Annex 1.
     <u>Official Journal of the European Communities</u>, L248,
     Volume 34 (5 September 1991).

39. J. Gaffney, A. Irsa, L. Friedman and E. Emken. <u>J.Agric.
     Fd. Chem.</u>, <u>27</u>, 475, 1979.

40. F.J. Winkler and H.L. Schmidt <u>Z.Lebensmitteluntersuch
     u. Forsch</u> <u>171</u>, 85, 1980

41. E.R. Schmid, H. Grundman and I. Fogy, <u>Ernahrung</u>, <u>10</u>,
     459, 1981.

42. A.R. Brause, J.M. Raterman, D.R. Petrus and
     L.W. Doner, <u>J. Ass. Off. Analyt. Chem.</u>, <u>67</u>, 5359, 1984.

43. A.R. Brause and J.M. Raterman, <u>J.Ass.Off.Analyt.Chem.</u>
     <u>65</u>, 846, 1982.

44. H.S. Lee & R.E. Wrolstad, <u>J.Ass.Off.Analyt.Chem.</u>,<u>71</u>,
     795, 1988.

45. J. Bricourt and J. Koziet, <u>J.Agric.Fd.Chem.</u>, <u>35</u>, 758,
     1987.

46. H. Craig, <u>Geochim.Cosmochim.Acta</u>, <u>12</u>, 133, 1957.

47. P.A. Guarino, <u>J.Ass.Off.Analyt.Chem.</u> <u>65</u>, 835, 1982.

48. B.A. McGaw, E. Milne & G.J. Duncan. <u>Biomedical and
     Environmental Mass Spectrometry</u>, <u>16</u>, 269, 1988.

49. Z. Sofer, <u>Analyt.Chem.</u>, <u>52</u>, 1389, 1980.

50. A.M. Pollard, <u>Chem Ind.</u>, 359, 1993.

51. J.B. Rossell, <u>Fat Science & Tech.</u>, <u>94</u>, 39, 1992.

52. J.B. Rossell, <u>Fat Science & Tech.</u>, <u>93</u>, 526, 1991.

53. K. Lee, P. Gillatt & J.B. Rossell, "Authenticity of
     Vegetable Oils and Fats Part XX. Determination of Maize
     Oil Purity by Stable Carbon Isotope Ratio Analysis"
     <u>Leatherhead Food RA Res. Rep.</u>, No. 719 (1994)[†]

†   Note. Leatherhead Food RA Reports are available on
      application to the Publications Department,
      Leatherhead Food Research Association, Randalls
      Road, Leatherhead, Surrey KT22 7RY, England.

Subject Index